GENETICS

PROJECTS FOR
YOUNG SCIENTISTS

PROJECTS FOR
YOUNG SCIENTISTS

GENETICS

BY MARTIN J.
GUTNIK

FINES ARE CHARGED

FRANKLIN WATTS
NEW YORK I LONDON I TORONTO I SYDNEY I 1985

DIAGRAMS BY
ANNE CANEVARI GREEN

Library of Congress Cataloging in Publication Data

Gutnik, Martin J.
Genetics.

(Projects for young scientists)
Includes index.
Summary: A collection of science projects which
demonstrate concepts and aspects of cell theory, cellular
reproduction, and modern genetic theory.
1. Genetics—Juvenile literature. (1. Genetics—
Experiments. 2. Experiments) I. Title. II. Series.
QH437.5.G88   1985   575.1'0076   85-8865
ISBN 0-531-15131-X (paper ed.)
ISBN 0-531-04936-1 (lib. bdg.)

CONTENTS

FOR
NATASHA

———

GENETICS

———

PROJECTS FOR
YOUNG SCIENTISTS

———

INTRODUCTION

The study of genetics is very complex. It has taken many years to reach our present level of comprehension. Genetics, like all sciences, is built with the bricks and mortar of past discoveries. It is to the early theoreticians that we owe a great debt of gratitude. Today, genetics is still a pioneer science, just beginning to reach toward its potential.

Genetics deals with the stuff life is made of and, like life itself, the field is highly complex.

The projects in this book will demonstrate the various aspects of cell theory; cellular reproduction, both asexual and sexual; concepts in Mendelian genetics; as well as many aspects of modern genetic theory. The projects will also attempt to knit together the ideas of these various areas, connecting thoughts and philosophies on modern genetic principles. In doing this, the author hopes to leave the reader with a clear understanding of today's genetic theories, how they were developed, and how they function.

All the experiments are designed to demonstrate cellular theory and past and present genetic principles. Many of these experiments will be quite complicated, and some potentially dangerous. *Special care must be taken by the experimenters.*

All of these experiments will be accompanied by explanations of the proper procedures for successful and safe performance.

1

WHAT IS GENETICS?

A golden eagle soars through the air and circles with the air currents, its keen eyes always alert for prey. Below a prairie dog uses its strong front paws to dig a hole in the ground into which it burrows for protection from the predator above. On a nearby hillside, a human being watches the drama unfold, capturing the events with a 35-millimeter camera.

Here, in a single environment, we have three different living creatures, three organisms, each unique both in behavior and biological makeup. What makes each special? What establishes the patterns that make them different?

Even among their own kind, the individuals in the scene above are unique. Perhaps this particular eagle prefers to soar for prey, while another eagle may perch and wait. The human being has brown hair and hazel eyes, and stands 5 feet 9 inches (175 cm) tall. How did this person develop these characteristics? Why do children often look like their parents, grandparents, or perhaps even like an aunt or uncle?

The answer to the above questions is, of course, heredity. *Heredity* is the passing of characteristics from one generation to the next. *Inheritance* is based upon heredity. Heredity is a process of life, and it is carried on by all living things. When we study heredity we are inquiring into inherited resemblances and differences.

Genetics is the analysis of heredity at work—the accounting for resemblances and differences. How, when, where, and why are some of the questions that geneticists attempt to answer. In essence, genetics is the science concerned with heredity and variation.

CELL THEORY

The study of heredity and genetics must begin with an understanding of a single cell. The *cell* is a unit of protoplasm or cytoplasm surrounded by a cell membrane. It is the unit of structure for all living things. Simply stated, *cell theory* holds that all living things are made up of cells. A single cell, in and of itself, is a living unit that carries on all the functions of living matter.

Cells vary in size. Most cells cannot be seen by the unaided eye; the tiniest cells, viruses and bacteria, can only be seen through a high-powered microscope. The largest single cell is the egg cell (yolk) of birds.

Cells vary in shape. There are as many varieties of cells as there are functions that an organism performs. A cell's shape and size help it to perform its function, whatever that may be.

Most cells duplicate themselves exactly. Cells, as stated above, are living things. All living things eventually die. If a cell did not duplicate itself, the organism to which it belonged would also die. Cells duplicate themselves through a process of cell division called *mitosis*.

Because cells are the "building blocks" of organic matter, they tend to organize into other structures. Cells organize into tissues, and tissues organize into organs and glands. A group of cells that all perform the same function is called a *tissue*. An *organ* is a group of tissues that perform similar functions. A *gland* is a group of tissues that perform similar functions and produce and secrete a liquid. Cells, tissues, organs, and glands form the systems that compose multicellular organisms.

It is heredity that keeps biological systems functioning. Heredity is the transmission of traits and characteristics in both *unicellular* (single-cell) and *multicellular* (many-celled) organisms.

2

WHAT IS A
SCIENCE PROJECT?

A science project involves the study of a problem or concept within any given scientific discipline. Genetic projects cover many areas—botany, zoology, cytology—as well as involving the fields of immunology, oncology, and evolution.

All science projects must have meaning to those performing them, as well as to others. Projects must follow the scientific methods of experiment and discovery, and they must be research-oriented.

SCIENTIFIC METHOD

All projects begin with the observation of scientific phenomena. *Observation* is using all your senses to find out all you can. Good observation techniques are the key to a successful project. All the other steps within the process of scientific research and discovery are based on the disciplined observations of the researcher.

After observing an event, a geneticist (or any other scientist) organizes the information that he or she has gathered. This is the *classification* of data. For example, *deoxyribonucleic acid, DNA*, has the ability to duplicate itself. All living things duplicate, or replicate, themselves. Therefore, all living things must possess DNA. Is DNA a property unique to living organisms?

The observations and resulting classifications of the scientific investigator should lead to a specific question or problem. This arises directly from observation and is called an inference or a prediction. An *inference* is an educated guess, based on what you have observed, about something that has happened. A *prediction* is an educated guess about something that is going to happen, based on what you have observed. The educated guess about the possible answer to a problem leads to the formulation of a hypothesis.

The *hypothesis* is the concept to be investigated. It is an inference or prediction that can be tested. It gives direction to the scientific investigation. In striving to prove or disprove a hypothesis, the investigator must always return to the basic premise. The investigation or experiment is based upon the hypothesis and relates directly to the stated problem. Your *results* will either confirm or refute your inference or prediction, leading to a *conclusion* that proves or disproves your hypothesis. If the hypothesis is wrong, it should be restated and tested again.

As you move through the scientific method of investigation, you will use many of the tools of science. Observations will be duly noted and recorded on tables of values, charts, or daily records. For example, you might keep a chart recording the process of mitosis in an onion cell.

By keeping the following chart, you have a ready, documented source upon which to base your classifications, inferences, predictions, and then your hypothesis. As you follow the proper procedures of the scientific method of investigation, you will minimize error and move from proper investigation to conclusion. Conclusions should be based on well-documented data only.

PLANT MITOSIS—ONION CELLS

DATE	PROCEDURES	OBSERVATIONS—VARIABLES
5/10/85 10:00 A.M.	Obtained prepared slides of onion root tip. All slides have more than one section.	The formation of two new cells from a mother cell takes approximately 20 hours, depending upon the given species and environmental conditions.

5/10/85 10:10 A.M.	Obtained slide, placed on microscope stage, and examined under medium and high power. Examined root tips for stained nuclei.	Most nuclei have been stained with one or more dyes. The nuclei have retained the dyes.
5/10/85 10:20 A.M.	Isolated one nucleus in which the chromatic material has formed a network of thread-like structures.	The chromatin seems to be longitudinally split and moving to opposite sides of the nucleus. The nuclear membrane seems to be breaking down, and the nucleolus has disappeared.
5/10/85 10:30 A.M.	Identified the above nucleus as entering the prophase of mitosis. Diagrammed and labeled cell.	

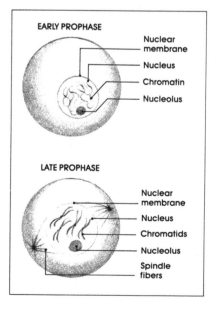

EARLY PROPHASE

Nuclear membrane

Nucleus

Chromatin

Nucleolus

LATE PROPHASE

Nuclear membrane

Nucleus

Chromatids

Nucleolus

Spindle fibers

GENETIC RESEARCH PROJECTS

Like all research projects, a genetics project must be based on sound scientific observation; for example, Francesco Redi's seventeenth-century experiments in *spontaneous generation*, in which he showed that "life must come from life."

The theory of spontaneous generation was that living things could arise from nonliving material. In the past, people believed this theory to be true. They were certain that maggots and other wormlike creatures were generated from decaying matter, that rats and mice gained life from rags, and that snakes came from horsehairs.

To modern thinkers this may all sound silly, but people not so very long ago did not have the knowledge that we do today. Science then was closely associated with black magic. Yet there were people who attempted to disprove the theory of spontaneous generation. Among them were Francesco Redi, Lazzaro Spallanzani, and Louis Pasteur.

Francesco Redi (1626?–98), an Italian physician, did not believe that maggots arose from decaying matter. He believed that life must come from life. He was the first to carry out controlled research projects in an attempt to disprove the spontaneous generation theory.

Redi cooked meat in order to kill any organisms that might be in it, and then put the meat into three separate, labeled containers. The first container was left uncovered and exposed to the air; the second was covered with a fine sterile gauze and sealed at the edges; and the third was completely sealed with heavy parchment.

Redi's theory was that the container exposed to the air, because of the odor and availability of the meat, would attract flies. The gauze-covered container, he guessed, would attract flies, because the odor would still escape and rise in the air, but the flies would not be able to enter the container. He believed that the parchment-sealed container would not attract flies because no odor would escape into the outside air.

This was a *controlled experiment*, or one in which the researcher keeps all experimental conditions constant except for the one factor being tested. Redi's hypothesis was that the meat would attract flies, and that the flies would lay eggs on the meat; then, after gestation, maggots would appear. This, he postulated, would prove that maggots come from flies and not from meat. Redi did not merely observe phenomena in his experiment, but he set up a controlled experiment in order to test for his hypothesized results. Most serious scientific research is accomplished by controlled experiments.

Redi observed the three containers over a period of time, logging the results of what happened in each one. This log reflected the progress of the experiment.

Redi observed that flies landed on the meat in the first container. They remained on the meat for a considerable period of time and then flew away. After several weeks, maggots appeared on the meat.

The flies landed on the gauze of the second container and remained there for quite some time. They could not enter the container to land on the meat. Several weeks later maggots appeared on the gauze covering of the container. There were no maggots on the meat.

For the duration of the experiment, flies did not attempt to land on the third, parchment-covered container. After several weeks there were no maggots on or in that container.

Redi repeated the experiment several times with the same results before coming to his conclusions. He concluded that, since only the first container had maggots on the meat, the maggots appeared as a result of the flies landing on the meat. This was also why there were maggots on the gauze of the second container and no maggots on or in the third. Thus, Redi reasoned, maggots come from flies, not from decaying material. Life comes from like forms of life.

Follow-up projects were performed by Lazzaro Spallanzani (1729–99) and Louis Pasteur (1822–95). Their experiments led them both also to conclude that life must come from life.

HOW TO APPLY
TO SCIENCE FAIRS

Many cities and towns conduct their own science fairs, often sponsored by the universities in the area. These institutions support science fairs in order to stimulate interest in science, as well as to encourage students to take up science careers. The fairs also help to discover scientific ability in the participants.

If you want to find out about entrance and eligibility requirements, get in touch with a nearby university, or check with the local library. Sometimes newspapers sponsor gifted applicants to science fairs, especially local ones. Awards at science fairs are usually given in the form of scholarships and trips.

All science fairs have their own rules and entry requirements. For information on the Westinghouse Science Talent Search and the International Fair for Science, write to Science Service, Inc., 1719 N Street NW, Washington, DC 20036.

Many organizations sponsor human genetics education centers that might be interested in a well-researched genetics project. Among the community organizations worth looking into are the Committee to Combat Huntington's Disease, the Association for Retarded Citizens, the Cystic Fibrosis Foundation, the March of Dimes Birth Defect Foundation, and the National Association for Sickle Cell Disease.

These genetics education centers were established in response to a need for improving the genetics content in school and college curricula. Some of their goals are to improve the study of human genetics, to provide contact with all professionals interested in human genetics education and study, to provide teachers and students with accurate and updated information about human genetics and its implications for the health and well-being of all members of the community, to provide teachers and students with information about the impact of genetic disorders and birth defects, and the development of programs to explore new ideas in human genetics and human genetics education.

Obtain the brochures and catalogs listing the guidelines for entrance and space reservations, and apply to science fairs early, since space may be limited.

IDENTIFYING A GENETICS PROJECT

Genetics is a broad field; almost anything within the discipline can become a stimulating science project.

Because genetics is such a varied science, it offers a great range of choice. You may wish to apply Mendel's ratios to Drosophila to confirm his findings. You could study human *genotypes* and *phenotypes* to demonstrate the roles played by dominant and recessive genes in human inheritance. You might want to do a project on biochemical genetics in peas. The possibilities are infinite.

Genetics projects should be research-oriented and include experiments or demonstrations to prove a specific theory or observation. For example, in mitosis and meiosis, do cells replicate themselves exactly, and what material in each specific cell triggers the reproductive process?

HOW TO SET UP AND PRESENT YOUR SCIENCE PROJECT

All research projects begin with general observations (goals) that the researcher hopes to prove (achieve). Suppose your science project is on cells. You might begin your project with the broad generalization: All things are made up of cells, and these cells are living units in and of themselves.

Cell theory states that all living things are composed of one or more units, called cells. All life functions are carried out by single cells or groups of cells. The life functions are: (1) ingestion of nutrients (digestion, absorption, and assimilation into the cell body); (2) movement (all living things must move); (3) growth; (4) change; (5) respiration; (6) reproduction (all living things must replenish their own kind); (7) excretion (expulsion of wastes); and (8) death (all living things must eventually die).

In multicellular organisms, the above functions are performed by tissues, organs, and glands.

MAIN OBSERVATION AND PREMISE
All cells come from preexisting cells. During cell division and reproduction, the determiners of hereditary characteristics are passed from parent to offspring.

The researcher now moves from his or her observations, or premise, toward classifying information and defining terms.

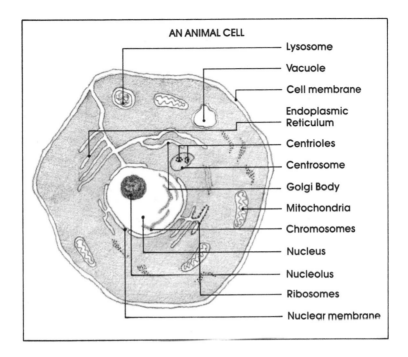

AN ANIMAL CELL

- Lysosome
- Vacuole
- Cell membrane
- Endoplasmic Reticulum
- Centrioles
- Centrosome
- Golgi Body
- Mitochondria
- Chromosomes
- Nucleus
- Nucleolus
- Ribosomes
- Nuclear membrane

Most animal and plant cells contain similar structures. The major difference between the two is that plant cells have a cell wall to give increased strength, shape, and form. Plant cells also have *plastids*, bodies that float in the liquid of the cell. For example, cells in green plants contain chloroplasts, which contain chlorophyll; animal cells do not have chloroplasts. Below, several of the major cell structures are defined.

The *nucleus* of a cell is a dense, spherical body of protoplasm, a suspension. Everything in the nucleus floats in the nucleoplasm (the liquid in the nucleus). The nucleus is the cellular structure that transmits genetic information in the form of DNA and chromosomes. The nucleus also regulates the life functions of a cell and takes an active role in, as well as controlling, cell division. The nucleus could be considered the brain of the cell. Nothing happens in a cell without direction from the nucleus.

The *nuclear membrane*, a thin layer of molecules,

separates the nucleus from the rest of the cell. It surrounds the nucleus and is selectively permeable, allowing for the passage of certain materials from the nucleus into the cytoplasm, and vice versa.

Within the nucleus, floating in the nucleoplasm, is the *nucleolus*, or little nucleus. This is a small, round body that is thought to transmit chemical information from the genes to the cytoplasm. It is composed of ribonucleic acid (RNA).

Chromatin, also found within the nucleus, is composed of loosely coiled, granular fibers floating in the nucleoplasm. When the cell starts to divide, the chromatin forms into chromosomes. *Chromosomes* are composed of deoxyribonucleic acid (DNA) and protein. They are rodlike structures that carry genes, which, in turn, transmit heritable traits. Chromatin comes from the Greek-derived word *chroma* (color), and takes its name from the fact that it stains darker than the rest of the cell.

Chromosomes are found within the nucleus of all cells (plant and animal), and occur in pairs. The genes on the chromosomes are concentrations of chromatin. A *gene* is a unit of heredity, which controls the development of a trait.

The *cytoplasm* in cells is protoplasm that surrounds the nucleus, and is bounded by the cell membrane. All structures within a cell are suspended in the cytoplasm and, on instructions from the cell nucleus, the cytoplasm performs most of the functions of the cell. The structures in the cytoplasm of the cell are called organelles. An *organelle* is a part of a cell that performs a specific function.

Vacuoles are spaces in the cytoplasm, often spherical in shape, and filled with a clear fluid, or *cell sap*. Plant cells usually have larger vacuoles than do animal cells.

The *cell membrane* is a thin, permeable surface that allows certain molecules to enter and others to leave the cytoplasm. Plant cells have a cell wall, thicker than the membrane, as well as the cell membrane itself.

Ribosomes are structures composed of ribonucleic acid (RNA) and protein, which function in the synthesis of protein. These structures are attached to the *endoplasmic reticulum*, a double membrane that forms a transport system within the cytoplasm.

The *centrosome*, a round or central body in animal cells, is found near the nucleus. It consists of two rodlike structures called *centrioles*. The centrioles function in cell division by aiding in the distribution of chromosomes. Surrounding the centrosome are fibers called *asters*.

The *mitochondria* are tiny structures scattered throughout the cytoplasm and bounded by a double membrane. They contain various enzymes that function in cellular respiration. For example: the Krebs cycle (oxidation) is carried out in the mitochondria.

Since the study of the cell also includes the examination of the process of cell division, researchers should include among their observations and classifications the structures that function in this process.

Chromosomes are the major carriers of genetic material, and possess certain properties that place them in a unique classification. These rodlike structures duplicate themselves precisely and divide equally during cell division. Chromosomes possess this capability because they are largely composed of a substance known as deoxyribonucleic acid (DNA).

DEVELOPING AN
INFERENCE OR PREDICTION

After observations have been classified, inferences and predictions can be made. Since the study of genetics is based upon the transmission of heritable traits from one generation to the next, you must first understand the processes of cellular reproduction—mitosis and meiosis—before you can deal with the study of genetics.

Developing an inference is based upon established patterns of learning. People must logically follow patterns in a series from beginning to end in order to comprehend the events that directly affect their lives.

MITOSIS

Cells grow by increasing their volume and outer surface membrane (see animal cell, p. 11). As the cell increases in surface area, its volume triples. Thus there is a larger increase in volume than there is in surface area. Practically speaking, the cell cannot grow fast enough to sustain its insides. There is not enough surface area to take in sufficient oxygen and food for the cell to survive. When a cell reaches this point, it must divide or die.

Mitosis is a process of cell division in which the resulting cells have the same chromosome complement as the original cell. Since two of the properties of life are growth and reproduction, all cells must experience mitosis or die. Mitosis is controlled by the nucleus of the cell. The nucleus is able to control this process because it contains genes (the determiners of inheritance). Genes are capable of making exact copies of themselves. In mitosis, when the nucleus divides, the chromosomes duplicate themselves and form two identical sets of genes (genes are on the chromosomes). The cytoplasm of the cell then divides, and a set of chromosomes goes to each new *daughter cell*.

The time required for mitotic division varies, depending upon the organism. An essential element of mitosis is that the two daughter cells are identical to the parent cell in chromosome complement and genetic information.

Interphase. Interphase is the period between divisions. It is the time for cellular growth, and lasts much longer than the division phases of mitosis. The cell grows until it reaches its maximum size. It must then either divide or die. Thus interphase ends in division or death.

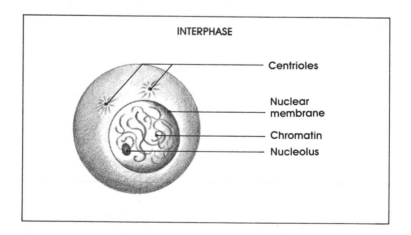

INTERPHASE

Centrioles

Nuclear membrane

Chromatin

Nucleolus

Prophase. During prophase, the cell prepares for the actual process of separating into distinguishable chromosomes. The chromosomes become extremely active, coiling as they condense into threadlike structures. After coiling, each chromosome is split lengthwise into a duplicate (sister) chromatid. These chromosomes split and move to opposite sides of the nucleus. The nucleolus disappears, and the nuclear membrane begins to break down. Mitosis has started.

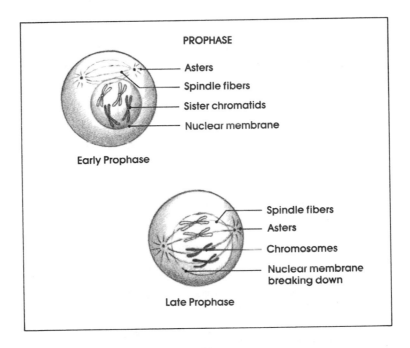

PROPHASE

Asters
Spindle fibers
Sister chromatids
Nuclear membrane

Early Prophase

Spindle fibers
Asters
Chromosomes
Nuclear membrane breaking down

Late Prophase

The genes have now duplicated themselves, and each chromosome has doubled lengthwise. The chromosomes become attached to the spindle fibers, forming two distinct sets of chromosomes. Spindle fibers are oblong and resemble a spindle. The end of the spindle is called a pole. The halfway point between the poles is called the equator. The spindle fibers are formed around the centrioles as the centrioles keep moving away from each other. This marks the end of prophase.

Metaphase. During metaphase there is a complete breakdown of the nuclear membrane. Paired chromatids

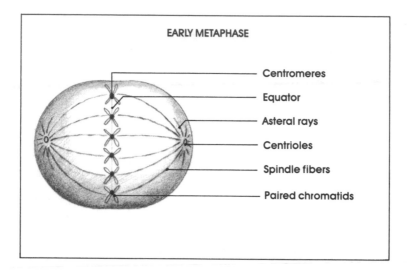

EARLY METAPHASE

- Centromeres
- Equator
- Asteral rays
- Centrioles
- Spindle fibers
- Paired chromatids

appear at the equator. The *centromeres* of each chromatid pair are attached at the equator. The asteral rays appear and form a network of spindle fibers.

Each chromosome becomes attached to a centrosome by a spindle fiber. These spindles are formed from the centrioles. Each duplicated chromosome now moves into a plane between the two poles of the spindles.

Anaphase. Anaphase can be considered the shortest stage in the mitotic process. During this phase each centromere divides, and each set of chromosomes moves away from its duplicate, until both sets reach the opposite poles of the cell. The centromere moves first, while the arms of the chromosome are dragged behind.

Telophase. The cell starts to revert to interphase state. At the poles the chromosomes in each set gather together and become less distinct. Once again they assume the appearance of chromatin threads. The spindle disperses. The nuclear membrane is reestablished and nuclei reappear. Finally cytoplasmic division is complete, and two daughter cells are formed. Daughter cells are two new cells, each approximately one-half the size of the mother cell. These cells will eventually grow to their maximum size and then divide.

Cytoplasmic division (cytokinesis) takes place when the cell membrane constricts during the telophase, and

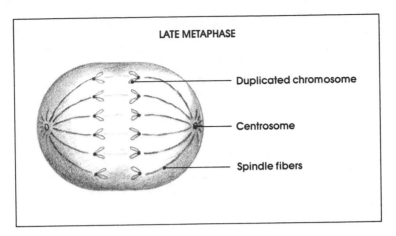

LATE METAPHASE

Duplicated chromosome

Centrosome

Spindle fibers

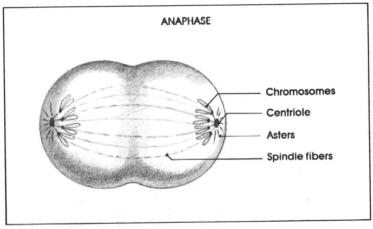

ANAPHASE

Chromosomes

Centriole

Asters

Spindle fibers

then divides the daughter nuclei. This results in one-half of the cytoplasm being distributed to each new cell.

MEIOSIS

Meiosis is the process of reduction division. During this process the sperm and the egg cells of most species reduce their number of chromosomes by one-half (the *haploid number*). This is a necessary step in sexual reproduction, for each individual species has its own specific chromosome number that is found in every cell. Human beings have twenty-three chromosome pairs, or forty-six chromosomes. If a sperm contains twenty-three pairs of

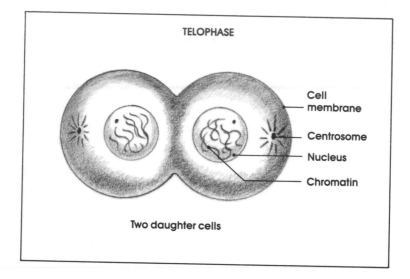

TELOPHASE

Cell membrane

Centrosome

Nucleus

Chromatin

Two daughter cells

chromosomes, and the egg also has twenty-three pairs of chromosomes, upon fertilization the embryo will have forty-six pairs, or ninety-two chromosomes. It is universally understood that this cannot happen if continuity of life is to be maintained.

Meiosis occurs during *spermatogenesis* (maturation of sperm cells) and *oogenesis* (maturation of egg cells), and results in one-half the number of chromosomes going to each mature sperm and egg cell. This process occurs in two separate nuclear divisions, called Meiosis I and Meiosis II. The first division, in which the chromosomes are halved, starts out similar to a mitotic division.

Meiosis I Prophase. The chromosomes appear as double threads within the nucleus, held together by a centromere. These chromatin threads become shorter and thicker, and the nucleolus disappears. As the nuclear membrane disappears, the spindle fibers become organized. Each chromosome pairs with its *homologue* (its match in size, shape, and gene order). The pairing of homologous chromosomes is called *synapsis.*

Meiosis I Metaphase. Synaptic pairs of chromosomes organize and line up midway between the poles of the spindle. Each synaptic pair consists of two chromosomes and four chromatids, and is associated with two spindle fibers. At the time of synapsis, some genetic material

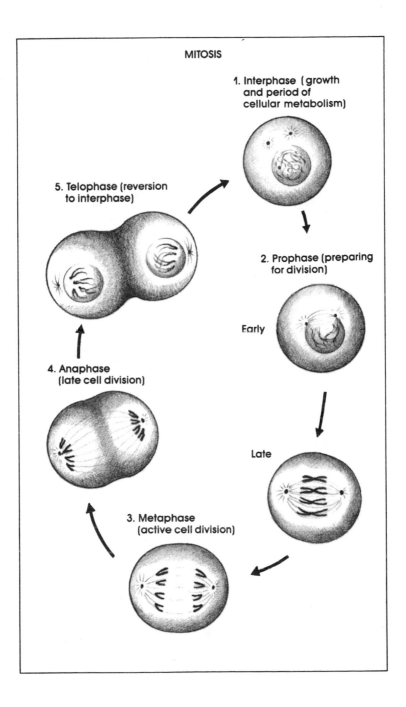

MITOSIS

1. Interphase (growth
and period of
cellular metabolism)

5. Telophase (reversion
to interphase)

2. Prophase (preparing
for division)

Early

4. Anaphase
(late cell division)

Late

3. Metaphase
(active cell division)

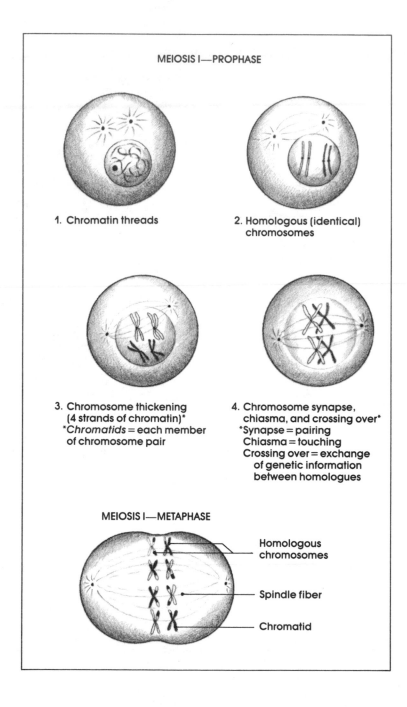

MEIOSIS I—PROPHASE

1. Chromatin threads

2. Homologous (identical) chromosomes

3. Chromosome thickening (4 strands of chromatin)*
   *Chromatids = each member of chromosome pair

4. Chromosome synapse, chiasma, and crossing over*
   *Synapse = pairing
   Chiasma = touching
   Crossing over = exchange of genetic information between homologues

MEIOSIS I—METAPHASE

Homologous chromosomes

Spindle fiber

Chromatid

may be exchanged between the homologous chromosomes. This exchange of material is known as *crossing over*.

Meiosis I Anaphase. During this stage, the chromosomes move toward the poles (ends of the spindle). In this process the centromeres do not split apart (in mitosis they do), and thus the homologous chromosomes become separated as they drift to the poles, but the sister chromatids of each homologue remain attached. Each daughter cell receives only one member of the homologous pair, or one-half the number of chromosomes.

Meiosis I Telophase. The original cell divides into two daughter cells, and nuclear membranes appear and surround the chromosome sets. Nucleoli appear, and the spindle fibers disappear.

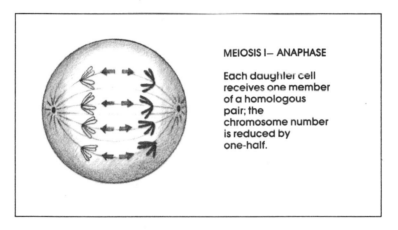

MEIOSIS I— ANAPHASE

Each daughter cell receives one member of a homologous pair; the chromosome number is reduced by one-half.

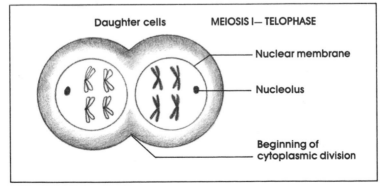

Daughter cells

MEIOSIS I— TELOPHASE

Nuclear membrane

Nucleolus

Beginning of cytoplasmic division

All the first stages of meiotic division are designed to reduce the number of chromosomes by one-half. Once this is accomplished, and after a short resting stage, the cells enter the second meiotic division.

Meiosis II. Meiosis occurs in two distinct stages. In Meiosis I, the homologous chromosomes of the parent cell are separated, and the chromosome number is halved in the daughter cells. Meiosis II is much like mitosis, but the end result is the formation of sex cells.

Meiosis results in the formation of gametes, each with a haploid number of chromosomes. For example, each human sperm cell formed has twenty-three single-stranded chromosomes. Each chromosome contains only one gene (not pairs of genes) for each trait represented on the specific chromosome. The gametes formed (both in the sperm and egg cells) contain varying genetic information due to crossing over.

The meiotic process is a necessary step for preservation of any given species. Both the male and female must go through this process, which is referred to as gametogenesis. The process is different in each sex.

FORMULATON OF
HYPOTHESIS OR STATEMENT
If cell division (mitosis) occurs in both plants and animals, then the phases of this process must be similar in both types of organisms.

TESTING THE HYPOTHESIS
(setting up an experimental plan)

MATERIALS FOR THE PROJECT
compound microscope
prepared slides of whitefish blastula
single-edged, safety razor blade
two blank slides
cover slips
fixing solution
acetocarmine dye
watchglass
toothpicks
alcohol burner
beaker or tumbler
fresh onion

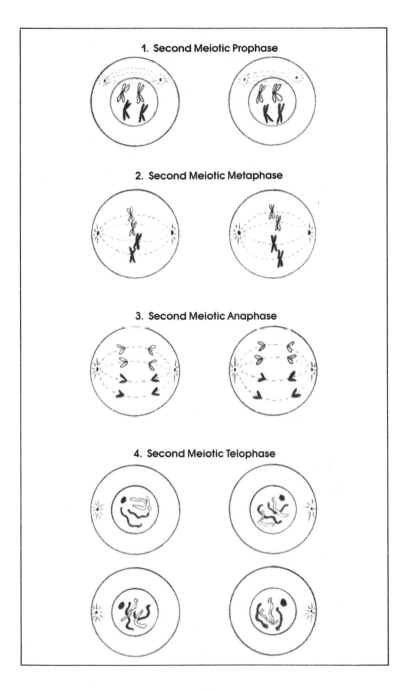

1. Second Meiotic Prophase

2. Second Meiotic Metaphase

3. Second Meiotic Anaphase

4. Second Meiotic Telophase

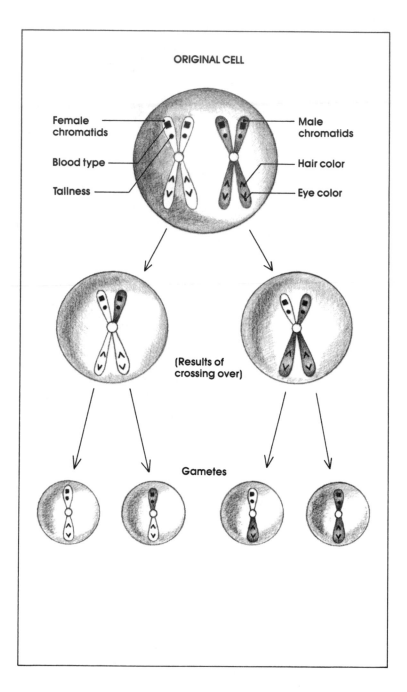

ORIGINAL CELL

Female chromatids

Blood type

Tallness

Male chromatids

Hair color

Eye color

(Results of crossing over)

Gametes

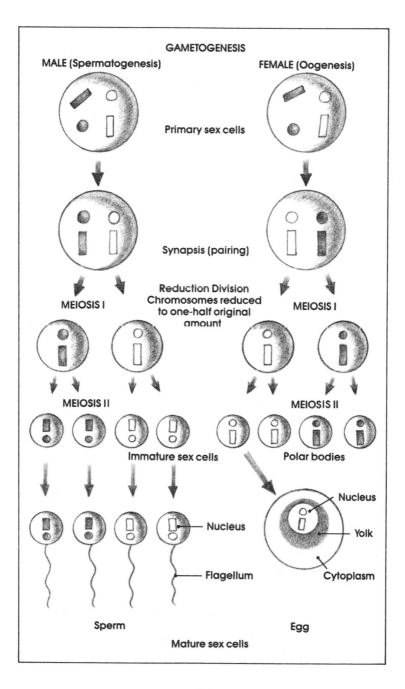

GAMETOGENESIS

MALE (Spermatogenesis)   FEMALE (Oogenesis)

Primary sex cells

Synapsis (pairing)

MEIOSIS I   Reduction Division   MEIOSIS I
                   Chromosomes reduced
                   to one-half original
                   amount

MEIOSIS II   MEIOSIS II

Immature sex cells   Polar bodies

                                               Nucleus

Nucleus   Yolk

Flagellum   Cytoplasm

Sperm   Egg

Mature sex cells

PROCEDURES
(PART ONE—ANIMAL MITOSIS)
Study the various mitotic phases (pp. 14–18). Examine each phase of the mitotic process, and learn how the cell evolves from one parent into two daughter cells. Examine the whitefish blastula under the compound microscope. Use low power. You should be able to observe cells in their various mitotic phases. Select a cell in anaphase.

Switch the microscope to high power, and observe the cell you selected. Can you see chromosomes and spindle fibers? Diagram the cell and label the chromosomes and spindle fibers.

Select a cell in telophase. Study this cell under high power. Do you see the cytoplasmic division? Diagram this cell and label the cytoplasmic division, chromosomes, and spindle fibers.

Compare your diagrams of the whitefish blastula to the similar stages shown on pages 17–19.

PROCEDURES
(PART TWO—PLANT MITOSIS)
The growth tissue in plants is called the *meristem*, which makes the plant grow in width and height. *Cambium* is meristematic tissue that is responsible for growth in the diameter of the stem. The main area of growth in most plants is the root tip. Here meristematic cells divide rapidly, making the plant grow taller.

Choose a fresh onion, and cut off the dried roots close to the base of the bulb. Fill a beaker or tumbler with water. Insert three toothpicks into the top of the onion to support it on the glass while its bottom is suspended in the water.

Put the suspended onion in a shady or dark place. Observe it daily. When the new roots reach ½ inch (1.27 cm), they are ready to be studied.

Fill a watchglass with acetocarmine dye. With a razor blade, slice off the lower part of the root tips and put them in the watchglass. Heat the root tips for approximately four or five minutes. Do not allow the stain to boil.

Remove a root tip from the stain and place it on a blank slide. Add one drop of acetocarmine dye. Observe which part of the root tip takes the darkest stain. Cut off

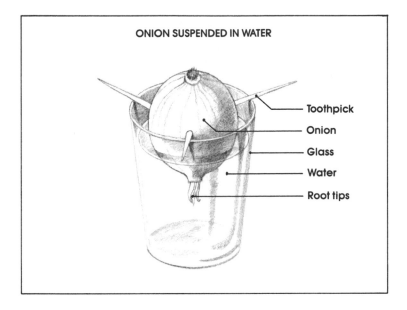

ONION SUSPENDED IN WATER

- Toothpick
- Onion
- Glass
- Water
- Root tips

this section, and chop it until the pieces are as tiny as possible. Put one of these pieces on a slide and cover it with a cover slip. Mash the tip until it is a thin layer on the slide.

Clean the slide, place it on the compound microscope, and examine the root tip under low power. You should be able to see cells in all the various stages of mitotic development.

Switch to high power and find a cell in anaphase. Diagram the cell and label the spindle fibers and chromosomes. Compare this diagram to your diagram of the whitefish blastula. Note the similarities and differences.

Find a cell in telophase. Diagram and label this cell. Label the cytoplasmic fissure, spindle fibers, and chromosomes. Compare this diagram with your diagram of the whitefish blastula in telophase. Note the similarities and differences.

ANALYZING RESULTS
AND OBSERVATIONS
The whitefish blastula and the onion root tip are representative of animal and plant mitosis. They clearly demon-

strate the similarities of the process in both the plant and animal kingdoms (refer to diagrams of whitefish blastula and onion root tips). Thus it can be inferred that mitosis in both plants and animals is the first step in passing on heritable characteristics (through the chromosomes) to future generations.

CONCLUSION
You conclude from your results that mitosis occurs in all organisms. Additional studies should be performed to confirm this conclusion.

You can further conclude, because mitosis occurs in all individual cells, that it is the first basic step in the genetic process.

PRESENTATION AT FAIR

Two displays of your project should be set up and labeled. For example, in the above experiment the slides under the microscopes, with your diagrams next to them, should be shown. The onion growing in water, diagrams of the various stages of mitosis, and examples of the onion root tip and the whitefish blastula in these various stages should also be presented.

Your charts on mitosis should be prominently displayed (usually on dividers or the walls of the project booth). These charts should be in color and well designed, and they should show the mitotic divisions in both plants and animals.

Your experiments involving whitefish blastula and onion root tips should be on two separate tables. (Of course this will vary, depending upon the space available.) Accompanying each experiment should be an easily readable explanation of procedure.

Above all, spectators and judges should have easy access to the booth.

THE FORMAL REPORT

A formal report of your project should be available for spectators and judges to read. You might have photocopies for distribution.

All science projects are more valuable and utilitarian when they are formally presented according to the standard practices of the given academic discipline. There are usually seven steps in devising such a report: observing, classifying, inferring or predicting, formulating a hypothesis, testing the hypothesis, analyzing results, and drawing conclusions. Your observations should be recorded and presented in a factual, objective manner. There is little room in science for subjective interpretation of phenomena. Once the basic research has been completed, all data must be classified and a specific problem identified and stated. This is an important step toward establishing the validity of your project. All scientists classify information. This lends order to the discipline, and puts information in its proper perspective for further research, investigation, or collection. The inference or prediction is based upon your research, and stems from your classification of data. These two abstractions are usually in the form of broad generalizations, reflecting the basic problems or theories of your report. The inference and prediction are the final tools utilized before arriving at the hypothesis.

The hypothesis is usually a statement reflecting the behavioral characteristics of your problem. It is a projection of your thoughts that can be validated by experimentation. Therefore, the development of an experiment to test your hypothesis must be formulated and carried out. All results during your experiment refer back to your hypothesis. Conclusions are drawn from the experiment's results. After the experiment is concluded, your formal write-up will read something like a recipe, allowing other researchers to use your discoveries and methods. Further research may be suggested or indicated, based on the identification of a new problem arising from your experiment.

3

MENDELIAN GENETICS AND
THE LAWS OF HEREDITY

Gregor Mendel (1822–84) is often referred to as the founder of genetic science. Mendel, an Austrian monk and practicing biologist, was the first person to study inherited traits, one at a time, and the first to propose theories and laws of heredity. His practice was to study single characteristics of organisms, rather than many factors at the same time. He also studied large populations of offspring, applying the mathematical theories of probability to his results.

Mendel worked mainly with garden pea plants, upon which he performed experiments, methodically recording his results. He published his findings in 1866, but they were not applied to experiments with chromosomes until approximately 1900.

Mendel's experiments led him to the development of certain theories. He tested these theories over and over before formulating his laws and principles. The laws and principles created by Mendel still stand today, basically unchanged.

Mendel developed two fundamental laws, or principles, to demonstrate the operation of heredity: the law of segregation and recombination and the law of independent assortment.

MENDELIAN TERMS

Mendel chose to work with garden peas because the plant was self-pollinating, and possessed traits that could easily be distinguished from one another—for example, color of flower or shape of seed. The monk cross-pollinated these plants for specific characteristics, and then studied them individually.

In order to experiment with and understand Mendelian principles, you must first know the following terms. Some of them may have been mentioned previously in this book, but they are worth mentioning again here.

Dominant character. The character that will always show up in first-generation offspring. The dominant character, when present, will always mask the presence of a recessive character. The dominant character is always represented by a capital letter.

Recessive character. Any character inherited by the offspring that does not show up (unless in pure form) in the first generation. Recessive characters usually show up in the second (F_2) and third (F_3) and successive generations. Their presence is always masked when a dominant character is present. The recessive character is always represented by a lowercase (small) letter.

Alleles. Genes that occupy corresponding positions on homologous chromosomes.

Homozygous. Having identical alleles for the same trait.

Heterozygous. Having dissimilar alleles for the same traits.

Hybrid. Any organism that inherits both the dominant and recessive characters from the parent.

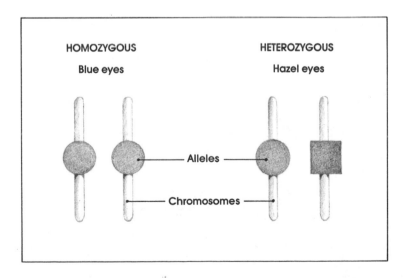

Genotype. The gene combination.

Phenotype. Physical appearance.

Law of dominance. After many experiments with successive generations, Mendel formulated this law of dominance: *When two organisms pure for contrasting characters are crossed, the dominant character appears in the hybrid, whereas the recessive character is hidden or masked.*

A *Punnett square* is used to demonstrate matings. Capital letters represent the dominant character, while lowercase letters represent recessive genes. F_1 stands for first-generation offspring; F_2 stands for second-generation offspring.

Let us consider, for our example, the mating of two individuals homozygous for the same trait.

B = Character for brown eyes (dominant)
b = Character for blue eyes (recessive)

	B	B
b	Bb	Bb
b	Bb	Bb

BB = father
bb = mother
F_1 = all Bb
(heterozygous hybrid)

As seen from the Punnett square, all the offspring in the F_1 generation will be brown-eyed hybrid. The dominant gene always shows up when present.

Mendel based his theory on crosses that he performed with the distinguishable characters in the garden pea plant. Using the Punnett square, let us examine some of the crosses that Mendel performed.

TALLNESS
T = gene for tall plants (dominant)
t = gene for short plants (recessive)
Genotypes: male = TT (pure tall)
female = TT (pure tall)

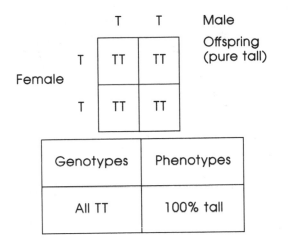

	T	T	Male
			Offspring (pure tall)
T	TT	TT	
T	TT	TT	

Female

Genotypes	Phenotypes
All TT	100% tall

As you can see, Mendel discovered that pure strains always breed true. Pure talls produce pure talls. Pure shorts will always breed pure shorts.

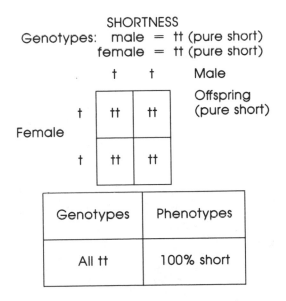

SHORTNESS
Genotypes: male = tt (pure short)
female = tt (pure short)

	t	t	Male
			Offspring (pure short)
t	tt	tt	
t	tt	tt	

Female

Genotypes	Phenotypes
All tt	100% short

When the recessive gene for shortness is pure in the genotypes, it will always show up in the phenotypes. Only when a dominant and recessive are present together is the recessive trait masked in the phenotypes (see below).

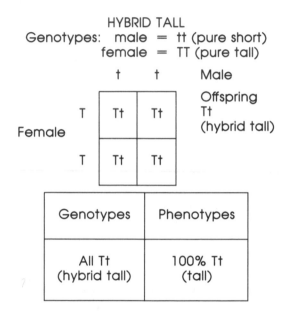

HYBRID TALL
Genotypes: male = tt (pure short)
                 female = TT (pure tall)

In the above example, the genes of the gametes mix, and we have one gamete for tallness and one for shortness. Yet, in the offspring, only the character for tallness (the dominant character) actually manifests itself. Dominant genes, by chemical action, overshadow recessive genes. The result of this overshadowing is a hybrid individual with contrasting genes for a trait. The individual is said to be heterozygous.

For example, let us look at the color of seeds in pea plants. Yellow is the dominant characteristic, and green is the recessive.

Genotypes: male = YY (pure yellow)
                 female = yy (pure green)

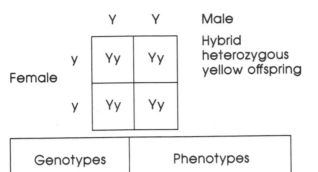

	Y	Y	Male
y	Yy	Yy	Hybrid heterozygous yellow offspring
y	Yy	Yy	

Genotypes	Phenotypes
All Yy (heterozygous hybrid)	100% yellow (dominant yellow overshadows green)

F_1 Generation

Law of segregation and recombination (Mendel's first law). The law of dominance, with all the offspring displaying the dominant character, holds true for the F_1 (first) generation. However, Mendel noticed that in the second generation (F_2) some of the offspring display the dominant characteristics while others display the recessive. Mendel also noticed that the number of offspring displaying the dominant characteristics is greater than the number of offspring displaying the recessive characteristics.

Because he was trained in mathematics, Mendel found this to be an interesting phenomenon. He inferred that this happened randomly and in predictable ratios. The offspring in the F_2 generation will segregate, phenotypically, into predictable ratios between dominant and recessive characters.

Formally stated: *If two hybrid organisms are mated, the recessive gene, which was masked in the first (F_1) generation, is segregated from the dominant gene. Upon a second fertilization, there is the possibility that these recessive genes will recombine and show up in the second (F_2) generation.* This is not one of Mendel's laws.

It is most desirable to demonstrate this concept with a *monohybrid cross*, where only a single pair of genes (characters) is involved. (The word *monohybrid* means having one pair of genes.)

PROBLEM
How do gametes segregate and recombine according to Mendel's first law?

MATERIALS FOR THE PROJECT
ear of corn with kernels all having a purple pericarp*
ear of corn with kernels all having a yellow pericarp*
ear of corn with homozygous kernels, pure purple and
    pure yellow (PP X pp)*
ear of corn with 100% heterozygous kernels, purple (Pp X
    Pp)*
calculator

HYPOTHESIS
If hybrid corn is crossed, then the resultant F_2 generations will segregate and recombine in a 3:1 ratio.

METHODS FOR THE PROJECT
Different varieties of corn are often distinguished by the color of the pericarp (fruit wall). Some varieties of corn have a purple pericarp, while others have yellow. We will assume that the purple pericarp is pure dominant, PP, and the yellow is pure recessive, pp.

    A cross is made (you have the crosses) to obtain the parental (F_1) generation. It may look like this:

PP = gene for purple (dominant)
pp = gene for yellow (recessive)
PP = male
pp = female

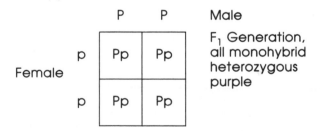

F_1 Generation, all monohybrid heterozygous purple

*May be obtained from most scientific supply houses

Genotypes	Phenotypes
All Pp	100% (purple)

Two plants from the F_1 generation are examined. Remember, these F_1 plants are the result of a cross between a pure homozygous purple and a pure homozygous yellow. Ask yourself the following question: Which parent does the offspring resemble? Why?

Study the ears of corn that represent the F_2 generation. Count the number of colored kernels on each ear, and record the numbers on a piece of paper. Calculate the ratio. What is the genotypic constitution of the F_2 generation? What is its phenotypic constitution? Use a Punnett square.

$$Pp = \text{male (heterozygous purple)}$$
$$Pp = \text{female (heterozygous purple)}$$

THE PUNNETT SQUARE

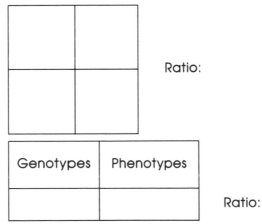

Ratio:

Genotypes	Phenotypes

Ratio:

PRESENTATION AT FAIR
Set up the booth according to the procedures mentioned previously in this book. The corn and charts should be displayed on your tables. Everything should be

labeled, and you should have one working example of your experiment and one hypothetical example. Your interpretation of the data should be clearly explained and displayed for the judges to examine.

Law of Independent Assortment (Mendel's second law). In order to study this law, it is necessary to perform a dihybrid cross. As you recall, in a monohybrid cross there is only a single trait and the organism is heterozygous for that trait. In a *dihybrid cross* there are two traits, each due to a pair of genes located on different pairs of chromosomes; the organism is heterozygous for both traits.

Mendel performed many crosses with dihybrid pea plants, and found that all the characters appeared (in the F_2 generation) in a 9:3:3:1 ratio. This is when he formulated his law of independent assortment: Each genetic trait behaves as an independent unit and is inherited independently of any other trait.

PROBLEM
How do traits behave as single units, and how are they inherited independently?

MATERIALS FOR THE PROJECT
ear of corn with kernels all having a purple pericarp
ear of corn with kernels all having a yellow pericarp
ear of corn with kernels crossed purple and yellow, 3:1 ratio
ear of corn with kernels crossed purple, starchy endosperm; yellow, sugary endosperm, 9:3:3:1 ratio
calculator

HYPOTHESIS
If a dihybrid cross always comes out to a 9:3:3:1 ratio, then genetic principles operate by predictable rules.

METHODS FOR THE PROJECT
This project will demonstrate how Mendelian ratios reflect the independent assortment of genes in a dihybrid cross. The presenter should use a booth with two tables; on one he/she can demonstrate a monohybrid cross, and on the other, the dihybrid cross. Charts and diagrams can be displayed on the booth's dividers.

Perform a monohybrid cross according to the procedures listed in the previous project. Show the results with a Punnett square and a genotype-phenotype chart. Ratios should be entered next to the charts.

A test cross is performed to determine if a plant of unknown genotype is homozygous or heterozygous. When you perform this type of cross, the offspring from the F_1 generation is mated with the homozygous recessive for the characters in question. For example, assume that the F_1 is all kernels with a purple pericarp. Take an F_1 kernel and mate it with a kernel from a yellow pericarp. The results should give you the genotype of the F_1 generation. If the results of this cross are all like the F_1 generation, you can assume the F_1 was homozygous dominant. If the results are a 1:1 ratio, you can assume the F_1 is heterozygous purple.

Cross 1

P P = F_1

	P	P
p	Pp	Pp
p	Pp	Pp

F_2

Cross 2

P P

	P	P
p	PP	pp
p	PP	pp

Genotypes	Phenotypes
All Pp heterozygous	100% purple

Genotypes	Phenotypes
50% PP heterozygous purple	50% purple
50% pp homozygous yellow	50% yellow

Dihybrid cross. You will not know what kind of corn plants were used to obtain your ear of corn. Count the characteristics (purple, yellow; starchy, nonstarchy). This should help you to determine the possible genotypes of the parents. Enter the count on a chart such as the one below. If you wish, do a test cross to confirm your findings.

TRAITS	NUMBER	TRAITS	NUMBER
Purple pericarp Yellow pericarp Starchy endo- sperm Sugary endo- sperm		purple—starchy yellow—starchy yellow—sugary purple—sugary	

Continue your study of the dihybrid cross by using your data to find the ratios for the F_2 generation. A Punnett square is an excellent tool for identifying the various genotypes of the F_2 generation.

PP = gene for purple
pp = gene for yellow
SS = gene for starchy
ss = gene for sugary

	PS	Ps	pS	ps
PS	PPSS	PPSs	PpSS	PpSs
Ps	PPSs	PPss	PpSs	Ppss
pS	PpSS	PpSs	ppSS	ppSs
ps	PpSs	Ppss	ppSs	ppss

Possible
male gametes

F_2 Generation
9:3:3:1

Possible female gametes

Interpretation. Of the above sixteen plants, nine will be purple with starchy endosperms, three will be yellow with starchy endosperms, three will be purple with sugary endosperms, and one will be yellow with a sugary endosperm. This gives us a 9:3:3:1 ratio for the phenotypic makeup of the F_2 generation. Because of this random combining and the ratios, we can now state that:

1. The genes for color and endosperms lie on separate chromosomes. They are separate units, inherited independently of one another.

2. The arrangement of chromosomes in pairing is due to chance, as is the union of gametes.

The specific ratios for purple, yellow, starchy, and sugary also fall into a predictable pattern. Use the chart on page 42 to show these ratios in genotypes and phenotypes.

There are twelve purple and four yellow plants—12/4 or a 3:1 ratio. There are twelve starchy and four sugary plants—12/4 or a 3:1 ratio. As you can see, the traits for color are inherited independently of those for endosperm. This is independent assortment.

Try this experiment with pea plants. Use tall (TT) and short (tt) pea plants, and yellow (YY) and green (yy) seeds. Follow the methods listed above. Do you obtain the same ratios? Does this verify your hypothesis?

PRESENTATION AT FAIR
Set up this project in a similar fashion to the previous one and, if you wish, demonstrate the two projects in conjunction. Once again, have well-documented displays, with charts and graphs on the dividers of the booth.

MENDEL AND BEYOND

By 1900 the Austrian monk's research had stimulated the scientific community. Using Mendel's work as a base, other scientists continued research into the field of genetics.

Walter S. Sutton, at the turn of the century, working with gametes under the microscope, concluded that the nuclear material in the sperm and the egg controlled heredity. He was the first to discover the process of meio-

Genotypes		Phenotypes	
1/16	PPSS; pure purple, pure starchy	1/16	purple; starchy endosperm
2/16	PPSs; pure purple, hybrid starchy	2/16	purple; starchy endosperm
2/16	PpSS; hybrid purple, pure starchy	2/16	purple; starchy endosperm
4/16	PpSs; hybrid purple, hybrid starchy	4/16	purple; starchy endosperm
1/16	PPss; pure purple, pure sugary	1/16	purple; sugary endosperm
2/16	Ppss; hybrid purple, pure sugary	2/16	purple; sugary endosperm
1/16	ppSS; pure yellow, pure starchy	1/16	yellow; starchy endosperm
2/16	ppSs; pure yellow, hybrid starchy	2/16	yellow; starchy endosperm
1/16	ppss; pure yellow, pure sugary	1/16	yellow; sugary endosperm

sis. He inferred that the sperm and egg must reduce their genetic material by one-half in order for the offspring to inherit an equal amount from each parent. (Mendel had said that the offspring inherit half of their traits from each parent.)

Studying chromosomes under the microscope, Sutton theorized on Mendel's law of independent assortment. He watched as the chromosomes, during meiosis, separated and went to the poles. He concluded that Mendel had been correct in his assumptions, and he further hypothesized that Mendel's traits were represented by tiny dots of genetic material, genes, that were located on the chromosomes.

Hugo de Vries (1848–1935), a Dutch botanist, formulated the concept of mutations. He inferred that changes in offspring were the result of changes in the genetic material. Through experiments with plants, he discovered that these changes were permanent. He therefore hypothesized that mutations were permanent changes in the genetic material.

Thomas Hunt Morgan (1866–1945), working with *Drosophila melanogaster* (fruit fly), further developed the concepts of dominant and recessive traits. He discovered a mutation in eye color, crossed the mutant with a normal, and discovered that the mutant gene was masked in the F_1 generation. When he crossed this first generation, some of the offspring in the F_2 generation manifested the mutant eye color. Morgan therefore hypothesized that the mutation was recessive, while the normal eye color was dominant.

Again using *Drosophila melanogaster*, Morgan did additional experiments with chromosomes. He studied crossing over and *sex linkage* and the positions of genes on the chromosomes. His experiments led to the chromosome theory of inheritance.

CODOMINANCE
(Incomplete Dominance)

Researchers who came after Mendel discovered that certain traits in offspring did not display dominant or recessive tendencies. Instead of looking like either parent, the offspring had an intermediate appearance. It seemed to these scientists that neither gene (the mother's or the father's) could control. In this case, the offspring was a true blend of both parents. Organisms that display lack of dominance for certain traits have at least three phenotypes in the F_2 generation, in a 1:2:1 ratio.

A good example of lack of dominance (codominance) lies in the crossing of red- and white-flowered four-o'clock plants.

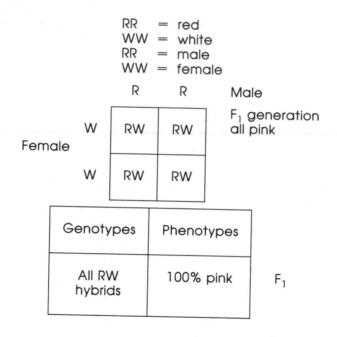

Genotypes	Phenotypes	
25% RR pure	25% red	
50% RW hybrid	50% pink	
25% WW pure	25% white	1:2:1 ratio

PROBLEM
How do blending traits show incomplete dominance?

MATERIALS FOR THE PROJECT
paper (graph paper, if desired)
pencil

HYPOTHESIS
If a trait displayed by both the male and female of a species is neither dominant nor recessive, then that trait will manifest itself in the offspring as a codominant.

METHODS FOR THE PROJECT
Let us assume that the gene for white fur in the ermine is neither dominant nor recessive (WW), and that the gene for black fur in the ermine is neither dominant nor recessive (BB). Perform a cross of a male white-furred ermine with a female black-furred ermine (WW $\times$ BB). What are the results for the F_1 generation? Show this in a Punnett square, and show the genotypes and phenotypes as done in the charts above. Are the offspring white, black, or gray? Why?

Now cross two offspring from the F_1 generation, a hybrid gray (WB) with a hybrid gray (WB). Show the results in a Punnett square and on a genotype-phenotype chart. What are the ratios of white to gray to black among the offspring?

Cross a homozygous white (WW) with a hybrid gray (WB). Show the results in a Punnett square and on a genotype-phenotype chart. What are the ratios of white to gray? Are there any black offspring? Why?

Perform this same type of procedure for Andalusian fowl, color BB × WW, neither being dominant or recessive; and for shorthorn cattle, RR = red, WW = white, RW = roan, neither red nor white being dominant.

PRESENTATION AT FAIR

Set up your booth in a similar fashion as described previously. If you wish, include this project as part of the project on Mendel's laws. Have your displays well labeled and documented.

4

PROJECTS ON THE PHYSICAL BASIS OF HEREDITY

Heredity is the transmission of genetic information from parents to offspring. This information is passed through the DNA molecules contained in the sperm and egg, and gives the gametes chemical instructions for the development of a new individual. The expression of this information in the offspring is the physical basis for heredity.

All organisms within the biosphere must, by virtue of being alive, conform to the eight life processes (see p. 10). One of these processes is death.

Since all living things must die, in order for a species to survive it must be able to replace itself. The process by which organisms replace themselves is called reproduction.

ASEXUAL REPRODUCTION

Asexual reproduction is the process of producing offspring from only one parent. Usually organisms that reproduce asexually lack sexual apparatus or possess nonfunctional sexual apparatus. In some organisms the sexual apparatus is nonfunctional only during asexual reproduction, while at other times in the organism's life cycle, it reproduces itself sexually. There are many varieties of asexual reproduction.

Binary fission. Much like the process of mitosis, where the nucleus and cytoplasm divide equally to form two new organisms, this type of reproductive process occurs in both simple plants and animals. Among the organisms that reproduce by binary fission are the amoeba and certain bacteria.

Budding. Many organisms, both plant and animal, reproduce by means of budding. In this process the nucleus divides equally, while the cytoplasm divides unequally. The main difference between budding and fission (see page 47) is that in fission the parent loses its identity to the two daughter cells, while in budding the parent keeps its identity. Yeasts, sponges, and hydras utilize budding.

Spore Formation. Those organisms that reproduce by budding and fission do so as the result of an outgrowth of tissues from the parent organism. However, in many multicellular organisms and in some unicellular organisms, special structures are present for reproduction. One type of specialized reproductive structure is the *sporangium,* more commonly referred to as a *spore case.*

The sporangium produces a large number of small cells called spores. Each *spore* has the capability to give rise to a new individual organism. When the spore case bursts, the spores are transported in the air or by water currents to new locations, where they will develop into adult organisms.

Vegetative Reproduction. The growth of a new plant from one of its parts is vegetative reproduction. For example, potatoes developing from the eye of a potato or a new shoot rising from a coleus cutting are types of vegetative reproduction.

Vegetative reproduction is part of the process of *regeneration,* the ability of a plant or animal to replace lost parts. Although all organisms utilize regeneration to some degree (people regenerate tissue to heal wounds, for example), some organisms have developed the process to the point of creating a completely new individual.

Some examples of vegetative reproduction are bulbs—short underground modified stems that swell and then develop into new individuals; tubers—similar to bulbs; runners—stems that grow along the ground and give rise to new organisms (strawberries utilize runners); and rhizomes—long stems that grow horizontally underground and give rise to new individuals (irises, ferns, and certain grasses reproduce by rhizomes).

An advantage of the asexual reproductive process is that the offspring are exact duplicates of the parents and therefore possess the exact characteristics of the original. Also, organisms that reproduce asexually produce large numbers of offspring within a short time period; this helps to ensure the survival of the species.

SEXUAL REPRODUCTION

Sexual reproduction is the process of two individuals giving rise to a new individual. The two individuals are referred to as the parents and the new individual as the offspring. The new organism begins with the union of two sex cells, one from the female and one from the male. Because it involves two cells from two different individuals, sexual reproduction inherently guarantees variability. This variability, in turn, has profound effects on heredity and the evolutionary process.

The union of a male and a female sex cell is called *fertilization.* The sex cells are referred to as *gametes.*

Fertilization, designed to guarantee the creation of a new individual, is a complicated process. Sperm, the male gametes, are usually liberated in tremendous numbers within the vicinity of the egg, the female gamete. Sperm cells are usually much smaller than the egg, and usually have some means of movement. The egg, on the other hand, is large in comparison to the sperm, and is usually not very mobile. The egg also carries a reserve food supply in order to nourish the developing individual once it is fertilized.

Sperm are attracted to the egg by chemical action. Usually, the first sperm to come in contact with the plasma membrane of an egg fertilizes that egg. Once a sperm enters an egg, the egg creates a chemical barrier to prevent other sperm cells from entering.

The nucleus of the sperm unites with the nucleus of the egg. At this moment—the union of the nuclei—fertilization occurs and a new individual immediately starts to develop. This new individual is called a *zygote.*

In some organisms, the zygote will remain dormant for a long period of time; in other organisms, it will begin to divide immediately. In any case, the zygote is the beginning of a new life. It will develop into an *embryo,* and the embryo into a new individual.

As a zygote develops into an embryo, and the embryo into an organism, it goes through rapid cell divisions (mitosis). Each cell divides and forms daughter cells.

These new cells organize into tissues, organs, and glands. Each cell in the developing individual possesses the genetic code inherited from the parents. The code is imprinted on the chromosomes by the genes. Thus it is absolutely necessary that each cell have the same number of chromosomes as the cells of the parents.

Referring back to *gametogenesis* (p. 25), you can see that both the sperm and egg cells go through the process of meiosis to reduce their chromosome number by one-half. A mature sperm cell possesses one-half the chromosomes found in each cell of the adult male, and a mature egg cell possesses one-half the chromosomes found in the cells of an adult female.

Upon fertilization—fusion of the nuclei of the sperm and egg—the chromosome number becomes equal to that in each cell of the parents. For example, each human cell contains twenty-three pairs of chromosomes, or forty-six chromosomes. In gametogenesis the chromosome number of both the sperm and the egg is reduced to twenty-three chromosomes, one of each pair. Thus, upon fusion, the chromosome number becomes twenty-three pairs again, or forty-six chromosomes. The offspring received one-half its chromosomes from the mother and one-half from the father.

This process of sexual reproduction and the mixing of chromosomes is a way of ensuring survival of the species. The offspring inherits traits from mother and father, and this combination usually makes for a stronger and more diverse individual. Adding to the possible variations is the process of crossing over (see p.21).

Crossing over helps to create the mixture of genes passed on to the offspring. Diversity is a concept recognized as necessary for survival in our natural environment. The more diverse an ecosystem (a natural environment), the better its chances for existence. This concept also holds true for individual organisms. Crossing over helps to create the gene mix necessary for survival.

SEXUAL REPRODUCTION
IN PLANTS
Most plants reproduce sexually. Some plants reproduce without flowers—thallophytes (spirogyra algae and oedogonium), bryophytes (mosses), and pteridophytes

(ferns)—and other sexually reproducing plants use flowers.

Angiosperms are plants that produce seeds within fruits. Since these types of plants (geranium, apple tree, raspberry, for example) are most familiar, the following description of their reproductive anatomy should be most useful in comprehending their sexual processes.

The flower is the reproductive organ of the angiosperm. Usually the flower contains both the male and female apparatus for gametogenesis. After the plant fertilizes the ova (seeds), the flowers disappear and fruits develop.

Sepals and *petals* are referred to as accessory reproductive organs. Their functions, although not directly related to the reproductive process, are to protect the reproductive organs and enhance the process.

Sepals are the outermost green leaflike structures located at the base of the flower. They serve to protect the flower as it is budding, vulnerable to frost and other environmental hazards. All the sepals as a unit are called the *calyx.*

The petals are the colorful part of the flower, sometimes containing glands that produce a nectar or scent. The purpose of the petals is to attract insects or birds to

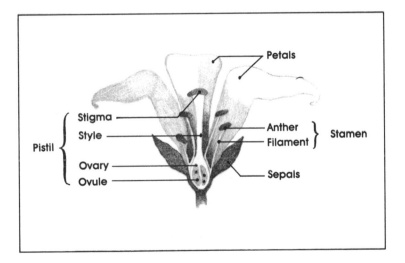

the plant so that they will help to pollinate (fertilize) the seed. The circle of petals is called the *corolla.*

The *stamen* (anther and filament) is the male reproductive organ of the angiosperm. Stamens are stalked structures that can be found within the corolla. Meiosis takes place within the anther, and pollen (sperm) is produced.

The *pistil* also has cells that undergo the meiotic process. The cells develop into an *ovule* (egg, or seed), the female reproductive cell. The pistil is designed to capture pollen grains and send them to the ovule so that fusion (fertilization) can take place. This is called *pollination.* In many angiosperms, *self-pollination* takes place. This means that pollen grains from the anther fertilize the ovule of the same flower. In other angiosperms, such as oak trees, *cross-pollination* occurs, and pollen grains from one flower are transported to the pistil of another flower on a different plant. Cross-pollination can be accomplished by the wind, and by birds, insects, and water.

The *stigma* of the pistil is sticky and thus specifically adapted to capture pollen grains. Once the pollen grains are captured by the stigma, they travel down the *style* (a hollow tube) to the ovary. The ovule is the female reproductive cell in the ovary.

Once the pollen reaches the ovule, its nucleus unites with the nucleus of the ovule and a zygote is formed. This zygote is called a seed. Immediately, mitosis ensues, and a new organism begins to develop. Eventually the zygote becomes the embryo of a new plant.

One of the requirements of life is movement (see p. 10). In order for a plant to survive, it must be able to disperse its seeds away from the parent plant. This will eliminate overcrowding around the parent plant, ensuring that certain individuals will grow into adult plants in a desirable environment.

Many plants have developed ingenious methods for the dispersal of their mature seeds. Many, like the maple tree, have seeds with wings, which can be carried by the wind in various directions away from the parent plant. Milkweed and cottonwood have fluffy, featherlike structures that can also be carried away by the wind.

Other plants, like the Canadian thistle or the cocklebur, have barbs or quills that attach to passing animals and are carried by them to different areas. Some seeds

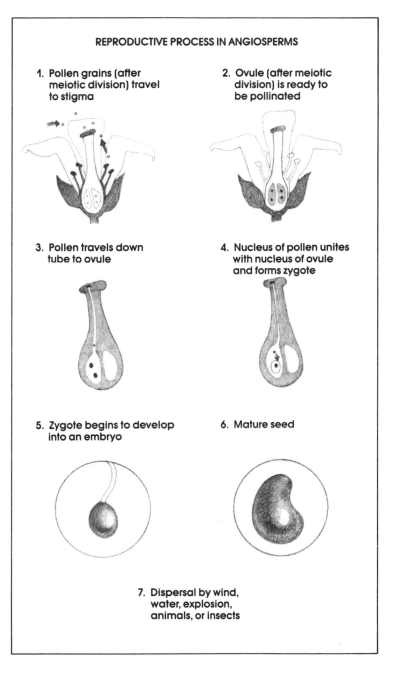

REPRODUCTIVE PROCESS IN ANGIOSPERMS

1. **Pollen grains (after meiotic division) travel to stigma**

2. **Ovule (after meiotic division) is ready to be pollinated**

3. **Pollen travels down tube to ovule**

4. **Nucleus of pollen unites with nucleus of ovule and forms zygote**

5. **Zygote begins to develop into an embryo**

6. **Mature seed**

7. **Dispersal by wind, water, explosion, animals, or insects**

float in water and are carried by rivers, streams, or runoff to different locations.

Some plants have fruits that explode and scatter their seeds in various directions. Finally, many plants have fruits that are eaten by birds or other animals; their seeds are later eliminated by the animals in different locations.

SEXUAL REPRODUCTION
IN ANIMALS

Most higher forms of animal life reproduce sexually and, like plants, these organisms have highly developed and specialized reproductive organs and glands. In the male of a species, the *testes* are the reproductive glands. The testes produce *sperm* (the male reproductive cells) and male *hormones.* In the female, the reproductive glands are the ovaries, which produce *ova* (eggs)—the female reproductive cells—and female hormones. The reproductive apparatuses are referred to as *gonads.*

Most animals, unlike plants, have reproductive glands of only one sex in each organism—the male in one and the female in the other. Often this is the only distinguishing characteristic between male and female. Some animals, such as earthworms, have both the male and female reproductive apparatus in one organism. These animals are referred to as *hermaphrodites.*

The nuclei of the sperm and egg cells are basically the same size. But the sperm cell is usually much smaller than the egg and is *motile* (capable of movement). The sperm contains very little cytoplasm and has a tail called a *flagellum*, which propels it to the egg. Males produce many more sperm than females produce eggs.

The egg is much larger than the sperm and does not exhibit movement. The egg contains a large amount of cytoplasm and, usually, stored food called a *yolk.* The egg cells of mammals, however, lack a yolk, and are much smaller than those of most other animals.

Fertilization. In order to create a new life, a sperm must fuse with an egg. All sperm are motile and must travel to the egg. However, the sperm contain very little cytoplasm and thus can only survive for a short period of time. Therefore both the sperm and the egg must be released in the same time period and in close proximity to one another or fertilization cannot take place.

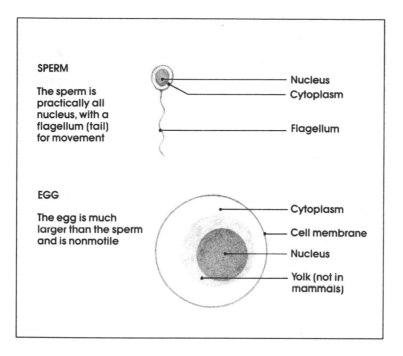

SPERM

The sperm is practically all nucleus, with a flagellum (tail) for movement

Nucleus
Cytoplasm

Flagellum

EGG

The egg is much larger than the sperm and is nonmotile

Cytoplasm

Cell membrane

Nucleus

Yolk (not in mammals)

The union of a sperm and egg is similar in all species. Usually, due to chemical action, the sperm is induced to swim toward the egg. Most sperm swim and must have a medium in which to do so. For example, fish sperm swim in water, and mammal sperm swim in *semen,* a liquid produced by the male. Among mammals, fertilization must be internal in order to allow the sperm cells to swim.

The sperm travels to the egg and, upon contact, its head penetrates the cell membrane and moves toward the nucleus of the egg cell. They fuse and form a single nucleus. Immediately the cells join and become a unit called a zygote.

Among some species an unfertilized ovum may give rise to a new individual. This is called *parthenogenesis.* This type of reproductive process is not uncommon among insects. Aphids, for example, reproduce in this manner. The offspring of parthenogenesis still have the *diploid number* of chromosomes—some because the egg did not go through meiosis, or some because two haploid eggs unite. However, in many insects the resulting offspring are usually males. These males, of course, do not exhibit many of the advantages of the sexual reproductive process. (Because there is no gene mix— from father and mother—the offspring do not display the increased vigor usually resulting from two parents with varying characteristics.)

THE HUMAN REPRODUCTIVE SYSTEM: THE BEGINNING OF HUMAN GENETICS

The reproductive systems of all species are specialized to produce offspring for continuation of that specific species. Unlike organs such as the brain, heart, or liver, these structures are not vital for the survival of the individual, but without them a species would soon become extinct.

The organs and glands of the reproductive system are concerned with the general process of reproduction. Some of these structures produce sex cells, some transport these cells, while others secrete hormones.

THE FEMALE HUMAN
REPRODUCTIVE SYSTEM
Anywhere from the ages of nine to fifteen, the female human may begin development from a girl into a woman. Upon a signal from the *pituitary gland*, the *ovaries* (the female sex glands) will begin to produce *estrogen* (the female hormone) and ripen ova.

The pituitary gland is the master endocrine gland (ductless gland), and controls all the other endocrine glands in the body. The ovaries, located in the lower abdomen, are activated through hormonal stimulation by the pituitary gland. Once they are stimulated, changes occur in the female body. The ovaries produce estrogen, which stimulates the following changes: The hips begin to broaden as the waist appears to narrow; the breasts begin to develop; the vocal cords constrict, raising the voice; hair develops under the arms and on the legs; pubic hair develops; and the ovaries begin to ripen ova (eggs). In the human female, one egg is usually ripened each month. One month the left ovary will ripen an egg, and the next month the right ovary will ripen an egg.

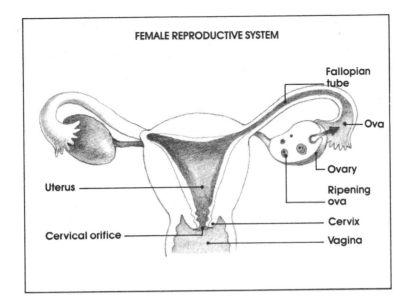

FEMALE REPRODUCTIVE SYSTEM

Fallopian tube

Ova

Ovary

Ripening ova

Cervix

Vagina

Uterus

Cervical orifice

The human female possesses all the reproductive apparatuses at birth, but it is not until puberty (adolescence) that these apparatuses begin to function. For example, at birth the ovaries contain approximately 200 eggs each. With the onset of puberty, the ovaries start to ripen these eggs, usually one egg every other month. Once an egg is ripened, it leaves the ovary and enters the *Fallopian tube.* This is called *ovulation.* The egg then travels through this tube to the *uterus.* The Fallopian tube is lined with hairlike structures called *cilia* that move the egg toward the uterus. It is while the egg is on this two-to-four-day journey between the ovary and the uterus that fertilization is most likely to occur. The egg is not fertilized in the ovary or the uterus—only in the Fallopian tube. Thus, the female is only fertile two to four days out of a month.

Most eggs reach the uterus unfertilized. While the egg is traveling, a chemical signal is transmitted to the uterus, alerting it to prepare for acceptance of the egg. In preparing, the uterus builds up a lining of blood to nourish the egg, should it be fertilized. When an unfertilized egg reaches the uterus, it imbeds itself on the opposite wall of the uterus and immediately begins to disintegrate. Approximately six to eight days later, the lining of blood and the disintegrated egg leave the uterus through the *vagina* (the female reproductive organ). This is called *menstruation.*

Menstruation occurs in a rhythmic fashion, monthly, from the onset of puberty. The menstrual cycle may be from twenty-five to thirty-one days, depending upon the individual. On a twenty-eight-day cycle, Day 1 is the start of *menses* (blood flowing). Days 13, 14, 15, and 16 are ovulation. Days 17 through 22, the egg travels to the wall of the uterus. Days 23 through 27, the egg disintegrates. On Day 28, menses begins again.

If the egg is fertilized (this must occur in the Fallopian tube), the nucleus immediately starts to divide, so that by the time the egg reaches the opposite wall of the uterus, the embryo is well on its way to development. The uterine blood lining joins with the embryo's tissue and becomes the *placenta*—through which the developing embryo receives nourishment from the mother via the umbilical cord.

THE MALE HUMAN
REPRODUCTIVE SYSTEM
Between the ages of ten and sixteen, the pituitary gland usually sends a signal to the two testes (the male reproductive glands) and, at that time, a boy will begin to develop into a man. Upon the signal from the pituitary gland, the testes will begin to produce *testosterone* (the male hormone) and sperm (the male reproductive cells).

With the production of the testosterone, changes will begin in the male body. Some of the changes will occur almost instantly, and some will develop over a period of years: The hips will narrow and the chest will broaden; the voice will lower; and chest, facial, underarm, and pubic hair will develop.

Unlike the ovaries, which each ripen only one egg every other month, the testes produce millions of sperm daily. The testes are located outside the body, in a sac called the *scrotum.* Sperm are most efficiently produced at a temperature of approximately 96°F (35°C). Since the human body temperature is usually 98.6°F (37°C), the testes are external to keep them two to three degrees cooler than the body.

As the testes produce sperm, they are stored in the thousands of tubules that make up the testes. In these tubules they mature and then move to the *epididymis* (a spongy layer of tubules that crowns each testis), where they are stored. The mature sperm later move into the *vasa deferentia* (tubes that carry sperm to the *seminal vesicles*), where they are stored, ready for *ejaculation* (expulsion from the body).

Upon sexual arousal, the *penis* becomes engorged with blood and becomes erect. The sperm then travel through the vasa deferentia to the seminal vesicles, where they are coated with an organic sugar in order to give them energy for their swim to the egg. From the seminal vesicles, the sperm travel into the *prostate gland*, where they become immersed in semen (the salty, or saline, liquid that sperm swim in). From the prostate gland, the sperm then travel into the *urethra* and out of the penis. This is called ejaculation.

Millions of sperm are ejaculated in order to ensure that some will reach the egg. (The human egg is approximately 200,000 times larger than the sperm.)

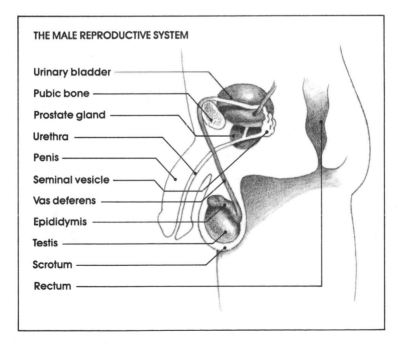

THE MALE REPRODUCTIVE SYSTEM

Urinary bladder

Pubic bone

Prostate gland

Urethra

Penis

Seminal vesicle

Vas deferens

Epididymis

Testis

Scrotum

Rectum

Fertilization. The sperm are ejaculated into the vagina and then must swim within the vagina, through the cervical orifice and into and through the length of the uterus, to the Fallopian tube that contains the traveling egg. Of the millions of ejaculated sperm, only a few hundred will reach the egg. These sperm will immediately surround the egg and attempt to break through its outer membrane. The first sperm to do this will be the one to fertilize the egg. The others will be prevented from entering by a chemical barrier produced by the egg.

The penetrating sperm will swim directly to the nucleus of the egg. Its nucleus will fuse with the nucleus of the egg, and a zygote will be formed. The egg has been fertilized. The egg immediately starts to divide, so that by the time it reaches the wall of the uterine sac it is an embryo.

The embryo will be enclosed by the embryonic membranes, surrounded by *amniotic fluid*, nourished by the placenta. In this warm, secure environment, the cells will divide and organize for nine months. The same genetic code will be in every cell of this new individual being.

GENETIC
CONTINUITY

Thus far this book has dealt with a variety of living species, and concepts that govern the life status of these organisms. It has been demonstrated that most living species have similar characteristics as far as the eight life processes are concerned. It has also been pointed out that each individual species of living organisms is unique, possessing individual characteristics that distinguish it from other species. This specialized information is passed on via the DNA molecules found in every cell of all living things. DNA provides the chemical instructions for inheritance and genetic continuity.

Inherited characteristics are those that are passed on from parent to offspring, generation after generation. Some of these *traits* are expressed in such visible characteristics as eye color, hair color, and height; others are expressed more subtly in blood types, intelligence, and immunity factors.

Acquired characteristics are physical variations influenced by environmental conditions. For example, suppose that identical twins grow up in two separate environments. One may develop in an affluent home, receive proper nutrition, and grow to be 6 feet (183 cm) tall. The other may grow up in a poor home, lack many of the nutritional needs, and only grow to be 5 feet 7 inches (170 cm) tall.

Two plants of the same species may be planted under different environmental conditions. One may be planted in rich soil with adequate moisture and sunlight. The other may be planted in poor soil with low moisture and inadequate sunlight. The former grows to be much taller and stronger than the latter, even though they both inherited the same potential. Environmental factors have a profound influence on the fulfillment of inherited characteristics.

Acquired characteristics cannot be inherited. However, though they cannot influence the following generation, they do tend to disrupt the present generation in its fulfillment of the genetic code. Acquired characteristics are variable, while inherited characteristics are stable. This stability is due to the composition of the DNA molecule.

DNA is a large, complex molecule, consisting of certain smaller chemical units that differ from one another. These smaller nucleotide units consist of deoxyribose (a type of sugar), a phosphoric acid, and four nitrogenous (nitrogen-containing) compounds. These four compounds are *adenine, thymine, guanine,* and *cytosine.*

In cells, DNA molecules assemble these smaller chemical units into other DNA molecules that are exact duplicates of themselves. This occurs during cell division. (For example, one of the reasons that skin cells can produce skin cells and brain cells can produce brain cells is that DNA—found in every cell—is self-duplicating.)

DNA takes the form of a twisted double chain, in which the nitrogenous compounds appear within the links. It is these combinations of nitrogenous compounds in the DNA chain that determine the properties of the molecule and the genes, which appear to be specific concentrations on the long molecule. In other words, if the nitrogenous compounds line up one way, a person may have brown eyes; if they line up another way, the eyes may be blue.

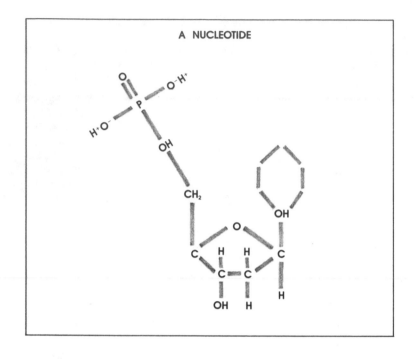

A NUCLEOTIDE

A DNA MOLECULE—
A DOUBLE HELIX

The DNA molecule is regularly organized into a double helix structure. It is two-stranded and coiled like a rope. The nitrogenous bases form steps between each of the strands. Each individual strand of DNA is like one-half of a spiral staircase. Each step on this staircase is made up of a pair of bases.

The biological significance of DNA is that it can replicate itself accurately during duplication. It has a very stable structure that rarely changes (mutates) during the process. Thus heritable traits remain the same. DNA has the potential to carry all kinds of biological information, and to transmit this information to the individual cell.

No two species or individuals possess DNA of exactly the same makeup, except identical twins. DNA is the code that determines much of the behavior and all of the features of a new individual (cell or organism).

Ribonucleic acid (RNA) is found mostly in the cyto-plasm, although some is found in the nucleus. RNA is

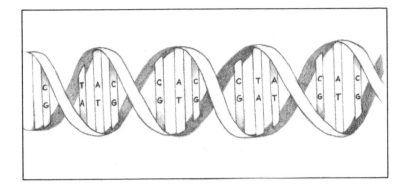

made of units composed mainly of ribose, a sugar similar to deoxyribose, but with an additional oxygen molecule. However, RNA replaces thymine with uracil and is only a single-stranded molecule. RNA aids DNA in carrying out its genetic and cellular activities.

In order for inherited characteristics to change from one generation to the next, there would have to be a change in the DNA of the sperm or egg cell. This very rarely happens, but when it does, the change is permanent. This type of change is called a mutation.

Mutations are changes that suddenly appear in an organism because of a change in the chromosomes or genes. The genetic information is permanently altered. Mutations can arise in a number of ways. Sometimes they occur spontaneously. At other times, they can be instigated by exposure to certain environmental factors, such as radiation or various chemicals. The organism in which the mutation occurs survives, and then passes this genetic change on to future generations.

Mutations in chromosomes are called chromosomal mutations, and aberrations of the genes are referred to as gene mutations. George W. Beadle and Edward L. Tatum, working with the mold Neurospora, discovered that mutations were due to the inability of a gene to produce any given enzyme produced by the parent plant. Because the gene was unable to produce a specific enzyme, it was incapable of performing its function.

Enzymes are composed of proteins. The four nucleotide bases in DNA are adenine, thymine, guanine, and cytosine. Uracil is found in RNA in place of thymine.

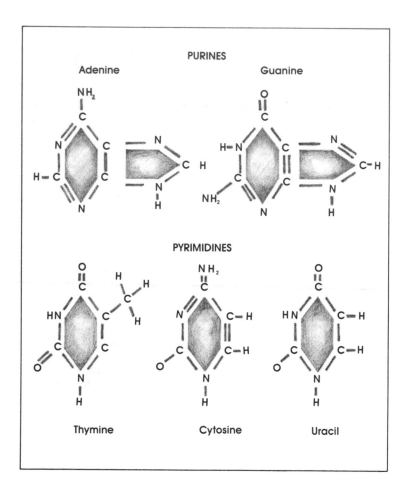

Purines (double-ringed) are adenine and guanine. Pyrimidines (single-ringed) are thymine, cytosine, and uracil.

The way purines and pyrimidines bond is very specific. The purine adenine will always bond with pyrimidines thymine or uracil. The pyrimidine cytosine will always bond with the purine guanine. This bonding is what ensures exact duplication in cell division. Because these substances can replicate themselves, they control the characteristics of organisms. Inheritance is therefore based on the order of bonding of the purine and pyrimidine bases.

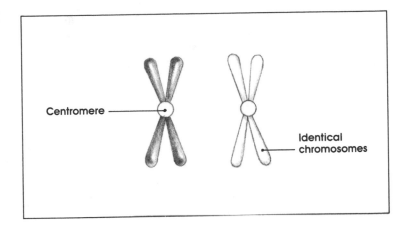

The phenomenon of crossing over of chromosomes occurs during the first meiotic metaphase. During this stage the chromatids are identical and held together by the centromere.

In prophase these homologous chromosomes approach each other, lie parallel to one another, and then become tightly intertwined. This is synapsis. It is during synapsis that the chromosomes may exchange parts and form chromatids with new combinations of genes. This new combination of information varies the genetic code in the offspring, making it different from the parents.

Remember that this is an exchange of information, not an altering of the genes. Crossing over is not a mutation. (It is important to understand that in crossing over, one chromosome comes from the mother and the other from the father; these two chromosomes make up the homologous pair. Thus, when crossing over occurs, the exchange is between paternal and maternal genes.)

Genes perform their functions by controlling the production of enzymes, which control the production of chemical substances, which, in turn, control the characteristics of organisms.

INFERENCES AND PREDICTIONS FOR
THE PHYSICAL BASIS OF HEREDITY
Since DNA is the only substance that can replicate itself exactly, and since chromosomes are composed of DNA,

then chromosomes must replicate themselves during sexual reproduction.

Since continuity is necessary for species reproduction, all species must have a specific unchangeable number of chromosomes.

PROBLEM
How can you prove that chromosomes carry heritable traits? Can you actually observe this phenomenon through experimentation? Does genetic continuity really exist?

MATERIALS FOR THE PROJECT
fruit flies (wild type)
culture jars (2-to-4 ounce (300 to 600 g) glass or plastic vials with porous stoppers—cotton is a good porous stopper)
anesthetizers
hand lens
dropper bottle of ether
morgue bottle (jar half-filled with mineral oil)
filter paper
camel-hair brush
Drosophila medium (may be obtained from most scientific supply houses)
cork and bottle
dowel
cotton
wire
electrical tape
white paper
jar cap (preferably from pickle or mayonnaise jar)

HYPOTHESIS
If chromosomes carry heritable traits, then manifestations of these traits will occur generation after generation, ensuring genetic continuity.

BACKGROUND FOR PROJECT:
DROSOPHILA MELANOGASTER
Drosophila melanogaster has been used in the study of genetics for many years. The Drosophila (common fruit fly) has only four pairs of chromosomes, which are large

and easy to study. Also, the fruit fly has a relatively short life cycle (10 to 14 days), which allows for short-duration experiments. Because the fly's life cycle is very short, so is its reproductive cycle. Large numbers of offspring with many contrasting traits are produced; because of the nature of the chromosomes, these traits are easily observable.

Like most insects, the Drosophila transmutates (changes form) in the process of metamorphosis. There are four individual stages in the Drosophila's development: egg, *larva, pupa,* and adult.

The larvae feed on the food medium, burrowing as they ingest the nutrient material. After two molts, the larvae leave the food supply and enter the pupa stage. During this stage, the pupae mature and darken in color. Finally, six days later, adult flies emerge and, within twelve hours, begin mating. The female flies then lay their eggs in the food supply. Gestation takes twenty-four hours, until the larvae emerge from the egg and begin the process once again.

All that is necessary to raise *Drosophila melanogaster* is to buy a culture medium (food supply) or make one, and then either to capture the adult flies in season or to purchase a small supply of them from a scientific supply house.

Once you have adult flies, the rest is quite simple. Keep the flies in an enclosed area (a small fish tank will do), with a food supply. The flies are very tiny, so a very narrow-mesh screen must be placed on the top of the tank to keep the flies inside.

Whenever you wish to study the flies, place a small vial of medium in the tank. The flies will go into the vial to lay their eggs. After the flies have landed on the medium, cover the vial with cotton and remove it from the tank. You now have your supply of Drosophila for study.

Anesthetizing Your Drosophila. In order to examine, count, and transfer your Drosophila, you must anesthetize them. The best method is etherization. *You must be very careful when working with ether. The fumes from this chemical can be hazardous to people. Also, ether is highly flammable. Do not use ether in a room with an open flame, and be sure to follow the directions on the bottle label.*

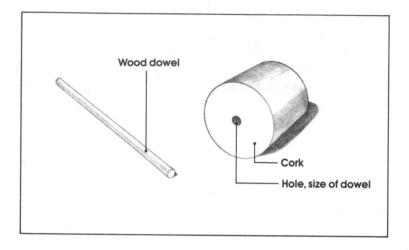

Wood dowel

Cork

Hole, size of dowel

You can purchase anesthetizing bottles from a scientific supply house. If you do not wish to purchase them, you can make your own.

Making an Anesthetizing Jar. You must have a bottle with a tight-fitting cork. Drill a hole the size of the dowel through the cork stopper.

Place the dowel through the hole you have drilled in the cork. Make sure that the dowel fits tightly. Place a piece of electrical tape over the top of the cork where the dowel has been inserted. This will ensure that the ether fumes cannot escape.

Test the cork and dowel to make certain they fit the bottle perfectly. Wrap several layers of absorbent cotton around the end of the dowel and secure it with a wire. Twist the wire very tightly so the cotton will not fall off the dowel.

Once you have captured enough flies in your vial, transfer them to the anesthetizing jar. Do this by placing the vial upside down in the jar. Place a few drops of ether on the cotton at the end of the dowel. Remove the cotton from the vial in the jar, and quickly place the cork over the jar.

The flies should be anesthetized in approximately ten seconds. Watch them carefully. After the flies have fallen to the bottom of the jar and are motionless, remove the cork from the jar.

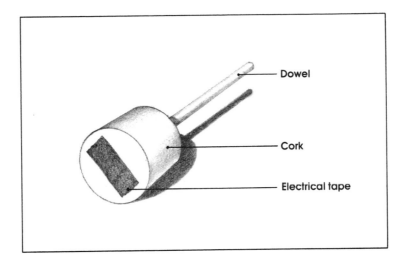

Dowel

Cork

Electrical tape

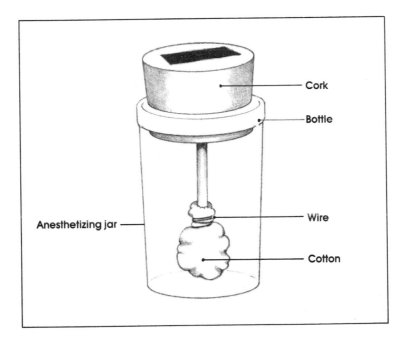

Cork

Bottle

Anesthetizing jar

Wire

Cotton

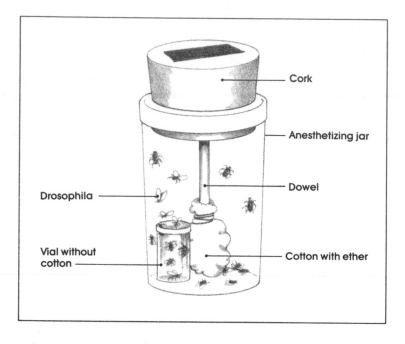

Cork

Anesthetizing jar

Dowel

Drosophila

Cotton with ether

Vial without cotton

Tap the bottle to remove the flies. Pour them onto a clean sheet of white paper. (Drosophila are easier to see on a white background.) With your hand lens, examine the flies to determine if any have died. Dead flies will have curled bodies and wings that stick out at an angle. Use your camel-hair brush to push the dead flies to one side. Put the dead flies in a morgue bottle.

The Drosophila that are still alive will remain anesthetized for approximately five to ten minutes. This gives you only a short time for sorting and study. If one of the flies you are examining begins to show signs of recovering from the anesthetic, reanesthetize it by using the following procedure: Cut a piece of filter paper to fit the inside of a pickle or mayonnaise jar cap. Place the filter paper inside the cap. Put a few drops of ether on the filter paper and cover the fly with the cap. This will reanesthetize the fly.

Identifying Characteristics. Sexual characteristics can most easily be identified by examination of the genital organs. The external reproductive organs of the Drosophila, both male and female, are located on the ventral posterior of the abdomen. The female has a slightly broader abdomen than the male, with small lines across the tip of the abdomen. The male organs are surrounded by heavy dark bristles, which give the posterior a blackish look. The male has a more rounded abdomen and sex combs on the fourth joint of the front legs. Males are smaller than females.

Long wings are dominant in wild-type fruit flies. Other wing types, such as vestigial, curved, bent, or miniature, are recessive.

Red eyes are dominant over brown eyes, white eyes, or black eyes. Gray is the normal dominant body color over yellow or black. Thus the normal wild-type Drosophila will have a gray body with long wings and red eyes.

METHODS FOR THE PROJECT
Place the fruit flies you have anesthetized on the white paper. Study these flies with a hand lens, and with the camel-hair brush separate them by sex. Put the males in

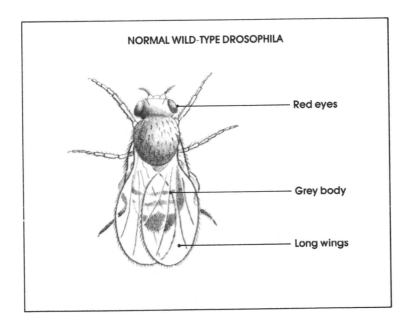

NORMAL WILD-TYPE DROSOPHILA

Red eyes

Grey body

Long wings

one culture jar and the females in another. Make sure you have selected normal males and females.

Now you are ready to cross the fruit flies to validate your hypothesis. Place three males and three females in the same culture jar and label it "Wild type—normal." Make sure there is enough medium for a good food supply. The flies should mate quickly and lay eggs within thirty-six hours. This first cross is the parent generation. Record the date of the cross.

Twelve days after the original cross the F_1 generation should emerge. Secure flies from the F_1 emergence and anesthetize them. Study them, looking for the characteristics described above. Record the results on a chart. (See p. 73.)

Secure six more flies from the F_1 generation and, using the same procedures described above, cross them. Chart the results of this cross. This will be the F_2 generation.

Are there any variations in the F_2 generation?

Continue your experiment by securing six wild-type males and six vestigial-winged females (vestigial-winged females can be ordered from a scientific supply house). Cross the males and females according to the procedures above. Chart your results. (See p. 74.)

Why do the offspring in the F_1 generation all have long wings? Because long wings are dominant. A dominant characteristic will always show up in the F_1 generation. (A Punnett square demonstrates this concept.) A dominant trait always manifests itself when present.

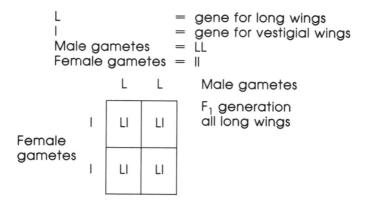

L   = gene for long wings
l   = gene for vestigial wings
Male gametes   = LL
Female gametes = ll

GENERATION	EYE COLOR	BODY COLOR	WING TYPE	SEX
parent	red	gray	long	M
parent	red	gray	long	M
parent	red	gray	long	M
parent	red	gray	long	F
parent	red	gray	long	F
parent	red	gray	long	F
F₁	red	gray	long	M
F₁	red	gray	long	M
F₁	red	gray	long	M
F₁	red	gray	long	F
F₁	red	gray	long	F
F₁	red	gray	long	F

GENERATION	EYE COLOR	BODY COLOR	WING TYPE	SEX
parent	red	gray	vestigial	F
parent	red	gray	vestigial	F
parent	red	gray	vestigial	F
parent	red	gray	long	M
parent	red	gray	long	M
parent	red	gray	long	M
F₁	red	gray	long	F
F₁	red	gray	long	F
F₁	red	gray	long	F
F₁	red	gray	long	M
F₁	red	gray	long	M
F₁	red	gray	long	M

GENERATION	EYE COLOR	BODY COLOR	WING TYPE	SEX
F_2	red	gray	long	F
F_2	red	gray	long	F
F_2	red	gray	long	F
F_2	red	gray	vestigial	M
F_2	red	gray	long	M
F_2	red	gray	long	M

What will happen in the F_2 generation? Using the previous procedures, mate two flies from the F_1 generation. Chart the results and demonstrate the concept with a Punnett square. (See p. 75.)

The F_2 generation will show a 3:1 ratio for vestigial wings. The Punnett square demonstrates the manifestations of the concept. The parents are hybrid (possess both dominant and recessive characteristics, but only the dominant characteristic manifests itself in the individual).

L = gene for long wings
l = gene for vestigial wings
Ll = male genotype
Ll = female genotype

	L	l	Male gametes
L	LL long	Ll long	F_2 generation 3 long wings and 1 vestigial— a 3:1 ratio
l	Ll long	ll vestigial	

Female gametes

As you can see from your charts and the Punnett squares, *Drosophila melanogaster* breed true, generation after generation. Thus you have proven your hypothesis to be correct and solved your basic problem.

PRESENTATION AT FAIR
You should have two sets of equipment displayed. One set should be a model of what you have done, and the other set of equipment should be functional. Accompanying your formal report should be samples of the tested Drosophila and copies of the charts and Punnett squares.

The tables should be set up in the front of the booth with any diagrams displayed on the dividers. Remember, the more organized and pleasing your presentation, the better your chances for recognition.

LINKAGE AND CROSSING OVER

Linkage is the tendency of genes to stay together. All chromosomes contain many genes, and these genes tend to stay together. All linked genes on chromosomes are aligned or paired in a specific order. This alignment may be broken when crossing over occurs.

A mutation is a change in the standard heredity pattern. When genes mutate, a new trait suddenly appears and is inherited. Mutations are usually harmful to the organism. For example, seedless oranges are harmful to the organism because this mutation does not allow the tree to reproduce. It is only through human intervention that seedless orange trees can be reproduced by grafting. (An organism that displays a mutation is called a mutant.)

Unexpected changes occur in organisms because of crossing over (see p. 21). Crossing over occurs during synapsis. During this stage of meiosis, the chromosome pairs sometimes twist and coil around one another. Then, upon separation, they exchange pieces (this is the crossover), and new combinations of characteristics appear in the offspring. Linkage is broken when crossing over occurs.

Since recessive mutant genes will not show up in the offspring unless they are two of the same inherited together, the chances for mutations becoming "visible" are very rare. But suppose there is one mutant gene on one chromosome and another mutant gene on its homologue (the other chromosome in the pair)? If this happens, when these homologous chromosomes split into gametes, each gamete receives one of them. As a result one could infer that the mutant genes could be passed on to different hybrid offspring. (Remember, a hybrid possesses both the dominant and recessive characteristics, but only the dominant manifests itself in the individual.)

There are mutations that occur on chromosomes other than the sex chromosome. These other chromosomes are called *autosomes.* When mutations occur among autosomes, they are referred to as autosomal dominant mutations, or autosomal abnormalities. This can happen during meiosis if a chromosome fails to separate. When a chromosome fails to separate, it is called *nondisjunction.*

If nondisjunction occurs, the cell may receive an extra chromosome. For example, if an egg cell has this extra chromosome and is fertilized, a condition occurs called *trisomy*. Down's syndrome is caused by trisomy resulting from an autosomal dominant mutation.

PROBLEM
Does crossing over cause recombinations in linked genes?

MATERIALS FOR
THE PROJECT
wild-type *Drosophila melanogaster*
brown (bw) vestigial-winged (vg) *Drosophila melanogaster*
bar-eyed (B) cut-wing (ct) *Drosophila melanogaster*
vermilion-eyed (v) scalloped-wing (sd) *Drosophila melanogaster*
culture vials (8)
anesthetizing and sorting equipment as in the previous experiment

HYPOTHESIS
If crossing over occurs in *Drosophila melanogaster,* then genes that are a farther distance apart are more likely to recombine than genes that are very close together.

METHODS FOR THE PROJECT
Obtain cultures of the above-listed strains of *Drosophila melanogaster* and mate them in the following order: (*a*) cross virgin wild-type, red eyes (R), long wings (L), gray body (G) with virgin brown (bw) vestigial-winged (vg); and (*b*) cross virgin bar-eyed (B) cut-wing (ct) with virgin vermilion-eyed (v) scalloped-wing (sd).

Cross approximately five or six pairs per culture vial, using four culture vials for each cross. It is wise to remove the parents from the culture vials after the cultures are fertilized. Also start to develop another batch of the mutant stock (all but wild-type Drosophila) so virgins will be ready to mate with the new F_1 generation.

In approximately ten to twelve days, the F_1 flies will emerge from the culture medium. Examine at least 200 of them and record their characteristics (phenotypes).

DATE	CROSS	PHENOTYPES
	Wild-type with brown vestigial wing	
	Bar-eyed cut-wing with vermilion-eyed cut-wing	

Continue with this experiment by making several cross-ings between the F₁ generation and another virgin pop-ulation of mutant stock. Then cross the F₂ generation with more mutant stock. Sort and list phenotypes. Study all phenotypes to observe which traits show up more often than others. Genes that lie close together have low fre-quencies of crossing over, therefore, among these genes there should be few recombinations. These genes are linked closely together. Genes that are farther apart on a chromosome recombine more often.

Observe the proportions of your recombinations, and create a linkage map by drawing a chromosome and labeling where each gene would lie on its surface.

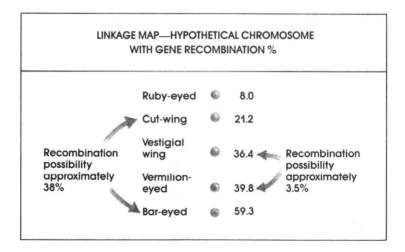

LINKAGE MAP—HYPOTHETICAL CHROMOSOME
WITH GENE RECOMBINATION %

Ruby-eyed 8.0
Cut-wing 21.2
Vestigial wing 36.4
Vermilion-eyed 39.8
Bar-eyed 59.3

Recombination possibility approximately 38%

Recombination possibility approximately 3.5%

PRESENTATION AT FAIR
This project can be demonstrated in the same manner as the previous project in this book. You might also choose to display this project and the previous project together to demonstrate how genetic continuity could be interrupted.

OTHER PROJECTS ON THE
PHYSICAL BASIS OF HEREDITY

Do a study on sex chromosomes and linkage. The sex chromosomes of the female are similar and referred to as the X chromosomes. In the male there is an X chromosome and a Y chromosome. In the human being, some genes, located on the sex chromosome, are for color vision. Attempt to show how this trait is linked to the sex chromosome.

5

HUMAN GENETICS

In many ways it is difficult to study heredity in human beings. Because people have a long life span and usually only produce one offspring at a time, you cannot cross them in a laboratory setting and record the results. The researcher may not live long enough to study the F_2 generation.

Human heredity is usually studied via family histories that record the transmission of heritable traits. Remember that the mechanisms of heredity ensure that the offspring will resemble the parent while, at the same time, also making that individual unique. The study of these similarities and differences is the study of human genetics.

Chromosomes, as you will recall, are found in the nuclei of all cells and contain genes, the determiners of individual traits. Each species of plant and animal has its own specific chromosome number. This number remains constant and the same for every cell in the organism. For people, that number is 46. Every cell of the human body has forty-six chromosomes, or twenty-three pairs. There are two exceptions to this rule: The gametes (sex cells) have only twenty-three chromosomes; and red blood cells, because they have no nuclei, have no chromosomes.

Thus a human zygote receives twenty-three chromosomes from its mother and twenty-three chromosomes from its father. Each cell of the developing embryo will contain twenty-three chromosome pairs. These sets of chromosomes that contain similar genetic material are called homologous pairs. In other words, a chromosome from the mother will pair up with a similar chromosome from the father. These sets possess similar genetic information (carry similar traits).

It is self-evident that if chromosomes are paired, the genes on these chromosomes must be similarly paired. These sets of similar genes are called alleles; one allele from the mother and one from the father. It is through the alleles that human inheritance manifests itself. Sometimes the information in a pair of alleles is almost exactly the same. For example, they both may contain information for blue eyes. The zygote inheriting this information is said to be homozygous (inherits the same genetic information from each parent).

When the information in the alleles is different—for example, one for brown eyes and one for blue eyes—the zygote is said to be heterozygous (the alleles differ).

Gene expression is the manifestation of the information in the alleles. Remember that the combination of genes in the zygote is the genotype, and the appearance of the individual is the phenotype.

SEX DETERMINATION

Autosomes are the chromosome pairs in the nucleus of the cell other than the sex chromosomes. The sex chromosomes are different from the rest of the chromosomes in the cell. In the female sex, the sex chromosomes are alike, but in the male sex, the chromosomes are different. The female sex chromosomes are designated XX; the male sex chromosomes are designated XY. The sex of a new individual is determined when fertilization occurs. The male determines the sex of the new individual. The zygote can only receive an X chromosome from the mother, but it may receive an X or Y chromosome from the father. If the zygote receives an X chromosome from the father, then the new individual will be a female, XX. If the zygote receives a Y chromosome from the father, then the new individual will be a male, XY.

POSSIBLE GAMETES

X = female
X = male

X or X = gametes from female
X or Y = gametes from male

	X	X	Female
X	XX	XX	
Y	XY	XY	

Male

Genotypes	Phenotypes
50% XX	50% female
50% XY	50% male

As you can see, the determination of sex is a 1:1 ratio. Ideally, for this to occur, you need a large population with a large number of fertilizations.

HUMAN PEDIGREES

A *pedigree* is the family history and the inheritance of certain traits. Genes are almost always either dominant or recessive. The inheritance of these dominant and recessive genes determines the human pedigree.

PROBLEM
What is the frequency of certain human genetic traits?

STATEMENT
Certain human traits manifest themselves more often than others.

METHODS FOR THE PROJECT
Each human trait is the result of the combination of genes inherited from each parent. You will examine two traits in yourself and in your family members to determine your pedigree.

The traits to be studied in this project are tongue rolling and earlobe attachment. The ability to roll one's tongue is a dominant characteristic, and is signified by the capital letter R. The inability to roll one's tongue is recessive, and is signified by the lowercase r.

In people the earlobes are either free or attached to the jaw. Free earlobes are dominant (L), and attached earlobes are recessive (l).

Interview and study the members of your family or, if you wish, your friends and members of their families. Check each person for tongue rolling and free or attached earlobes. Record the information on a pedigree chart (see example below).

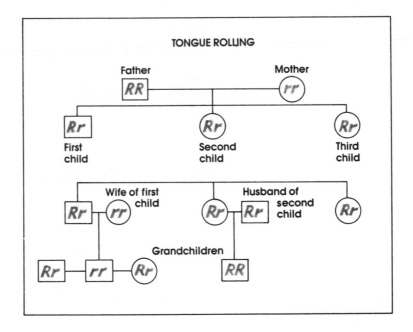

You may also show the results of your surveys in a Punnett square and genotype-phenotype chart. This will demonstrate the possibilities for each child and the ratios that can be expected.

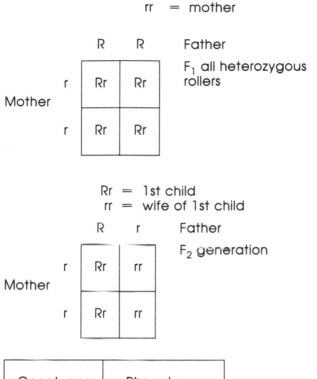

RR = father
rr = mother

R   R   Father

F₁ all heterozygous rollers

Mother

	R	R
r	Rr	Rr
r	Rr	Rr

Rr = 1st child
rr = wife of 1st child

R   r   Father

F_2 generation

Mother

	R	r
r	Rr	rr
r	Rr	rr

Genotypes	Phenotypes
50% Rr	50% rollers
50% rr	50% nonrollers

F_2 generation
1:1 ratio

Make a pedigree chart for each trait you study, and answer the following questions on the basis of your results: Is the gene (trait) you studied dominant or recessive? Could two parents who are rollers have a nonroller child? (Ask the same questions in regard to free or attached earlobes.)

PRESENTATION AT FAIR

Set up your project booth as described earlier in this book. Have your pedigrees clearly labeled and prominently displayed. Use Punnett squares and genotype-phenotype charts to explain the probabilities and ratios. You may wish to take this project further by having visitors to your booth fill out information cards on themselves, referring to the traits being studied, so that you can make pedigrees for these people as well.

SEX LINKAGE

Certain genes located on the X chromosome are responsible for certain traits that are not related to sexual development. Because these genes are on the X chromosome, they have no partner on the Y chromosome. Thus, if the zygote is a male (XY), these traits are more likely to be expressed than if the zygote is a female (XX).

Sex-linked disorders occur more frequently in males than in females. This is because the genes, whether dominant or recessive, appear only on the X chromosome and have no partner on the Y chromosome. Because of this, they will always be expressed. The Y chromosome apparently lacks genes for sex-linked traits.

Researchers seem to agree that this sex linkage is due to the different sizes of the X and Y chromosomes. Because the Y chromosome is much smaller than the X chromosome, the alleles for sex-linked traits do not appear on the Y chromosome. It is because of this that recessive traits are expressed in the male offspring.

Color blindness is a common sex-linked trait, as is hemophilia. A color-blind male is usually unable to distinguish between greens and reds. Hemophilia is a condition in which the blood does not clot normally.

The above conditions are more common in males than in females. The recessive gene that causes these maladies is inherited by the male from his mother and then passed on to his daughters. For example, assume that a female with normal vision marries a color-blind male. Let (C) designate normal vision and (c) designate color blindness.

$$X^C X^C = \text{female genotype}$$
$$X^c Y^- = \text{male genotype}$$

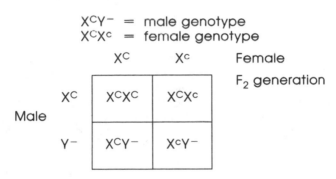

	X^C	X^C	Female
X^c	X^CX^c	X^CX^c	F_1 generation
Y^-	X^CY^-	X^CY^-	

Male (left label)

Genotypes	Phenotypes
50% X^CX^c female carriers	50% Normal color-vision females
50% X^CY^- 50% color-blind males	50% normal males

Let us assume that one of the carrier females, X^CX^c, marries a normal male, X^CY^-.

$$X^CY^- = \text{male genotype}$$
$$X^CX^c = \text{female genotype}$$

	X^C	X^c	Female
X^C	X^CX^C	X^CX^c	F_2 generation
Y^-	X^CY^-	X^cY^-	

Male (left label)

As you can see from the chart on p. 88, the color-blind male passed the recessive gene to one of his daughters, and when that daughter had children, she gave the color-blind gene to a son. If you wish, do a Punnett square to see what happens when a color-blind son marries a female carrier. You may also study the other sex-linked traits mentioned above with a Punnett square.

Genotypes	Phenotypes
25% X^CX^C	25% normal females
25% X^CX^c	25% carrier females
25% X^CY^-	25% normal males
25% X^cY^-	25% color-blind males

PROBLEM
How to determine a sex-linked pedigree.

STATEMENT
Sex linkage is passed from mother to son and from that son to his daughters. The important feature regarding sex linkage is no father-to-son transmission.

METHODS FOR PROJECT
Perform a pedigree study of yourself and your family to see what sex-linked characteristics you or your siblings may have inherited. Interview all the family members and record the results on a pedigree chart. Of the four sex-linked characteristics mentioned earlier, are any displayed in your family? From your results, can you reaffirm the statement for this study?

PRESENTATION AT FAIR
Use the same methods as in the previous project on human pedigrees.

SEX-LINKED INHERITANCE IN DROSOPHILA MELANOGASTER

Since it is impossible to study human beings in a laboratory setting, sex linkage in Drosophila melanogaster can be studied instead. In Drosophila there is one set of sex chromosomes. As in humans, both the male and female possess an X chromosome of about the same size and shape. But the male fruit fly, like the human male, does not have a second X chromosome, but rather a smaller, hooked, Y chromosome.

PROBLEM
What happens to the genes in *Drosophila melanogaster* when they are sex linked?

MATERIALS FOR THE PROJECT
stock of *Drosophila melanogaster*, white eye color
stock of *Drosophila melanogaster*, wild type
culture jars (4)
Drosophila melanogaster materials as listed on p. 66.

HYPOTHESIS
If genes are sex linked in *Drosophila melanogaster*, then the mothers will pass those genes on to the sons and the sons to their daughters.

METHODS FOR THE PROJECT
Using the identification methods listed on pp. 70–71, find and cross three white-eyed virgin females with three wild-type males. Put these insects in a separate culture jar and label it properly. Mate three wild-type virgin females with three white-eyed males. Put these in a separate culture jar and label it. Wait approximately eight to nine days and observe the culture jars to see if the eggs are developing. If so, remove the adults.

The F_1 generation flies should emerge in ten to twelve days. Remove approximately sixty to seventy from each culture and study them for eye color and sex. Prepare a chart to record your information (see example below).

F_1 GENERATION
DROSOPHILA MELANOGASTER

White-eyed	Sex	Red-eyed	Sex

Take three white-eyed females from the F_2 generation and put them in a fresh culture jar with three red-eyed males from the same generation. Do a cross with white-eyed males and red-eyed females. Follow the same procedure as above, and chart the results for approximately 100 of the F_2 generation offspring.

Ask yourself the following questions: Which eye color is dominant? Are either white or red eye color carried as an allele on the Y chromosome? Do a Punnett square to demonstrate the results of your experiment and a genotype-phenotype chart. Use the X and Y to represent the sex chromosomes and superscripts for eye color. For example, if you found that red eye color was dominant and carried as an allele on the Y chromosome, you would write them as follows: $X^R Y^R$ = red-eyed male; $X^r X^r$ = white-eyed female. If you found white eye color to be dominant, but lacking an allele on the Y chromosome, you would write it as follows: $X^W X^W$ = white-eyed female; $X^W Y^-$ = white-eyed male; or, if white is recessive: $X^w X^w$ = white-eyed female; $X^w Y^-$ = white-eyed male. Remember that if white eyes are recessive, they will show up in sex linkage in the male because there is no expression of that gene on the Y chromosome.

What genotypes must exist for white-eyed females? Do the mothers pass a sex-linked gene to the sons? Do the sons pass this on to their daughters?

PRESENTATION AT FAIR

This project may be presented as a generalized statement on sex linkage. Note that sex linkage in human beings works essentially in the same manner. You may wish to present this entire section on sex linkage as one project at the fair. It should be displayed as suggested in previous projects. You may want to include well-researched photographs (examples of sex linkage in humans) with descriptive captions.

OTHER PROJECTS IN HUMAN GENETICS

Eugenics. Eugenics is a method of improving the human condition through the improvement of heredity. Some methods for practicing eugenics are screening people for negative characteristics, such as mental illness and physical disease. The problem with eugenics is that in order to prevent these characteristics from being passed on, the carriers must be prevented from reproducing. However, the methods for preventing reproduction are unacceptable to most people (for example, sterilization and institutionalization).

Do a study of the history of eugenics and attempt to find a method of bettering the human condition through heredity. Remember that your solutions should be stated in positive, not negative, form.

Euthenics. Euthenics is a method for improving the human condition through the improvement of the environment. Since the moral issues attached to a eugenics program may prohibit improving the human situation, a euthenics program seems to be much more utilitarian.

Many human conditions, such as poverty, lack of education, and urban blight, need to be corrected. Do a study of the human condition as it relates to its surroundings. Attempt to devise a method for improving these conditions, thus effecting a positive change for the future.

GENETICS AND
THE ENVIRONMENT

Are children prodigies because of the genes they inherited from their parents, or because of the environment in which they are raised? What gives one person a gift for art, another a mind for mathematics, and yet another perfect pitch? Geneticists may argue that such characteristics are all inherited, while environmentalists may say that they are acquired and then formed. Who is correct?

It *is* true that certain human characteristics are inherited and unaffected by the environment. For example, eye color, attached or unattached earlobes, and blood type are all inherited characteristics that are not affected by environmental conditions.

Most human characteristics can, however, be influenced by the environment. In essence, most people are molded by a mixture of hereditary and environmental conditions. A child may inherit the potential for musical genius, but if that child is not exposed to music, that potential will never be developed.

Another person may inherit the potential to become 6 feet (183 cm) tall, yet, because of a nutritionally deprived environment, that person may only grow to be 5 feet 8 inches (173 cm) tall. Skin color, although inherited, can be changed by the environment, and some diseases that cause bone deformations can be attributed to poor diet.

It is apparent that although most of our characteristics are inherited, we are what we are because of many influences. The total human being is a blend of hereditary, environmental, and educational factors working together either in accord or in a discordant fashion.

Do a study of how heredity, environment, and education have molded the people in your family. Use charts and graphs to demonstrate the influence that one factor has had upon the other. Show how each individual is a mixture of these influences. From your research, project how these factors may be altered for improvement.

6

GENETICS AND ORGANIC EVOLUTION

Are people descended from apes? Did the giraffe's neck evolve into its present length or was it always long? Why do we have an appendix if it does not really function? Are we evolving today? What will we evolve into? If DNA is stable, how does evolution occur?

These are some of the questions you may ask yourself about the subject of evolution. *Evolution* is the process of genetic change that occurs in all matter. The process is slow, and it involves the interaction of substances with their organic and inorganic environments.

"Organic evolution" refers to the evolving of living material. According to scientific theory, all life began with the development of protoplasm into a single cell. Over time, this single cell developed specializations and eventually transmuted (changed in form) into multicellular organisms. These organisms in turn developed specializations, each reacting in individual fashion to its environment, each becoming more complex. It is this process—the change from simplicity to complexity, from nonvaried to varied—that is referred to as organic evolution.

How do scientists determine that evolution really took place? What is their evidence? What are their theories?

THEORIES OF EVOLUTION

Any ecologist will tell you that there is a continuity within the biosphere; that all living things interrelate with their living and nonliving environments; and that life itself is a continuous, changing process. What is the key to success in this process? Survival! Evolution, the process of change, is what survival is all about.

The study of fossils tells us that all organisms seem to have descended from a simple, common ancestor—the single cell. This cell transmuted in order to specialize itself for survival. Thus complex species developed from simpler species. These more complex species interacted with their environments and responded to ecological conditions through adaptation. Those that could not adapt became extinct. But how did the different organisms on Earth adapt? Why did they survive while others failed? What genetic properties allowed them to conform to their environmental conditions?

The process of heredity is usually stable, passing characteristics for survival from one generation to the next. This stability is due to the nature and character of DNA itself. (Remember that DNA is the only living substance that is capable of replication.) Because of its nature, the composition of DNA is usually constant. It is this consistency that contributes to the stability of the species.

Yet, once in a while, changes do occur in DNA. These changes occur in the genes of a gamete and are then passed on to the offspring. This new characteristic is called a mutation (a new inherited characteristic). The individual, or offspring, that inherits this characteristic is called a mutant. Usually mutants do not survive. The change is normally harmful to the organism. However, some mutations do help an organism to adapt to environmental conditions. It is these mutations that give rise to the evolutionary process. *In order for a change to be evolutionary, it must be able to be inherited.* If the change cannot be passed from generation to generation, it will die out with the individual that possesses it.

Jean de Lamarck (1744–1829), a French naturalist, theorized that the environment caused organisms to develop certain structures. If such a structure was useful in the process of survival, then it became stronger and larger. If it was not useful, it would eventually disappear.

In essence, Lamarck was referring to acquired characteristics—for example, the calluses that develop on a construction worker's palms. Lamarck insisted that these acquired characteristics are inherited and further developed by future generations. Yet August Weismann (1834–1914), in an experiment in which he cut the tails off mice and then mated them, disproved Lamarck's theory. The tailless mice gave birth to mice with tails.

Charles Darwin (1809–82), a British naturalist, developed his theory of evolution while serving as science officer on the Beagle—a ship of the British Navy that visited the Galapagos Islands off the western coast of South America.

Darwin collected and studied numerous samples of plant and animal life. As he studied his notes, he began to see patterns of similarity. He pondered what made the finches of the Galapagos Islands so similar while each individual species was so different.

In Darwin's time theories had already been published that the Earth had evolved over many millions of years; that rocks, landforms, and continents had changed. Darwin postulated that if nonliving things could evolve over time, then so could living things.

Darwin next examined fossilized evidence of evolution. Through his survey of animal remains, he found more and more evidence to support his theories. For example, geological surveys presented evidence that supported the theory that the distribution of species over the surface of the globe was directly related to the past distribution of these species' possible ancestors.

Darwin then studied geological records for evidence of the variance of structures among individuals in a species. Over the years, he found enough evidence to infer that changes occur due to individual variations within a species. According to his research, he found that the individuals that possessed structures more suited to their environments survived, while the others perished. Those that survived passed these adaptations on to the next generation.

Finally, after many years of research, Darwin published his theories: The Origin of Species by Means of Natural Selection. In this work he stated that the species of plants and animals that populate the present-day biosphere are direct, though modified, descendants of species that existed many years ago. According to Darwin, these present-day species evolved to their current state through the process of natural selection.

TODAY'S THEORY
OF EVOLUTION

Lamarck's theory of use and disuse, Darwin's theory of natural selection, and De Vries's theory of mutation have led us to the modern theory of evolution. This theory, based on the research of the past, utilizes the major points of these perceptive researchers, but corrects any errors they made due to lack of knowledge. The theory states:

> All living species reproduce their own kind. For survival purposes, they tend to produce many more offspring than can survive. These offspring struggle for existence and, through natural selection, only the fittest survive.
>
> The organisms that are most likely to survive are those that possess variations most suited to their environment. These variations (mutations) occur naturally. Some of them are harmful to the individual organisms. Such organisms will usually not reach adulthood. Other variations are useful and help the individuals to adapt more readily than their competitors. Such organisms will survive and pass the traits on to future generations. This is known as survival of the fittest, or best adapted.
>
> Mutations that are beneficial to the individual organism are genetically passed to succeeding generations. Acquired characteristics are not passed on. Organisms such as insects, which can genetically adapt themselves quickly to changing environmental conditions, are the most successful.

As you can see, heredity plays an important role in the evolutionary process. Changes in individuals that occur through dominance, recessiveness, mutations, and crossing over are indelible manifestations in the offspring, and are passed on from generation to generation. If these hereditary changes help the organism to survive, the individuals with the changes will continue to exist, while those without them will become extinct.

EVOLUTION OF VERTEBRATES

A fossil is the naturally preserved remains of an organism that lived ages ago. A fossil may be the actual organism itself or parts of that organism, or an imprint made by that organism on a rock or a shell, or in the soil. For years scientists have studied fossil remains to determine the patterns of the evolutionary process.

Living organisms can also be studied for evolutionary data. By studying the organisms, links can be established between the higher and lower vertebrates. By studying vertebrates, you should be able to establish evidence that shows the relationships between each species.

PROBLEM
Are there structural relationships between the higher and lower vertebrates?

MATERIALS FOR THE PROJECT
a dog
a frog
a turtle
a canary or parakeet
photos of the internal skeleton of each
a guppy or goldfish
water
paper
pencil

HYPOTHESIS
If there are relationships beteen vertebrates, then their body structures should be somewhat similar.

METHODS FOR THE PROJECT
Make a chart with the five classes of vertebrates you are studying. A *vertebrate* is any animal with a backbone (the backbone is part of the endoskeleton, or internal skeleton).

Vertebrates are among the most complex organisms on Earth. Their bodies, beside the endoskeleton, are composed of highly sophisticated organ systems and nervous systems.

In this project, five classes of vertebrates will be studied.

1. *Fishes*—class, *Pisces*. Some characteristics are cold-bloodedness (body temperature alters with temperature of environment), internal gills (these extract dissolved oxygen from water), fins, streamlined bodies, air bladders (these allow a fish to raise and lower itself in the water), and an internal bony skeleton.

2. *Amphibians*—class, *Amphibia*. The amphibian is also a cold-blooded animal. The body has a smooth, shiny, hairless skin. Amphibians spend part of their life in water and part on land. They lay eggs in water and these eggs hatch into tadpoles. Tadpoles live in the water until their bodies, internally and externally, alter into the adult form. This is called metamorphosis. Amphibia have an internal bony skeleton.

3. *Reptiles*—class, *Reptilia*. Reptiles are cold-blooded, their bodies are covered with dry scales, they breathe by means of lungs, and most reptiles lay eggs. All reptiles have an internal bony skeleton.

4. *Birds*—class, *Aves*. Birds, unlike reptiles and amphibians, are warm-blooded (body temperature always the same—constant). However, unlike reptiles and amphibians, birds have four-chambered hearts. Reptiles and amphibians have three-chambered hearts. A bird's body is covered with feathers, and it has an internal bony skeleton.

5. *Mammals*—class, *Mammalia*. Mammals are also warm-blooded, and their bodies are covered with hair or fur. Their babies are born alive and are nourished by mother's milk, produced by mammary glands. Their bodies possess an internal muscular wall, the diaphragm, which is the major breathing muscle. All mammals have an internal bony skeleton.

Through research and study, compare the structures of the goldfish, frog, turtle, parakeet, and dog.

Make a chart similar to the one on page 99. Fill in the chart by the actual studies of these animals and by research (photographs, books, and films). Answer the following questions:

1. What are the major similarities among your five classes of vertebrates? Differences?

2. Do you see any evidence of interrelationships among these species?

CHART ON VERTEBRATES

STRUCTURE	GOLDFISH (PISCES)	FROG (AMPHIBIA)	TURTLE (REPTILIA)	PARAKEET (AVES)	DOG (MAMMALIA)
1. Heart					
2. Skeleton					
3. Blood					
4. Reproduction					
5. Breathing apparatus					
6. Body covering					
7. Foot structure					
8. Tail					
9. Jaw					
10. Specific adaptations					

3. From your study and research, can you conclude that these species descended from a common ancestor?

PRESENTATION AT FAIR
Set up your project booth as described for earlier projects in this book. Have your charts, photographs, and other research well documented and labeled. If you use live animals as part of your display, make sure they are in a secure cage to ensure their safety as well as that of the spectators.

OTHER PROJECTS ON EVOLUTION

Do a project to see what effect natural selection has on the occurrence of particular alleles in a population. This can be accomplished by studying gene frequencies in given populations, and then studying their changes by the process of natural selection.

Study the evolution of the giraffe's neck and present the process according to Lamarck's theory of use and disuse, Darwin's theory of natural selection, De Vries's theory of mutation, and the modern theory of evolution.

7

MEDICINE AND
MODERN GENETICS

Modern medicine has reached a point where most researchers agree that many of the diseases we know are directly related to their genetic composition. Some researchers go so far as to say that all disease is genetic in origin. They come to this conclusion because the immune system is inherited. Since disease is controlled by a person's immune system, a person who inherits a weak immune system is subject to diseases that a person with a stronger immune system can resist.

Many diseases today are considered specifically genetic in origin; for example, *Tay-Sachs disease, sickle cell anemia,* and *Down's syndrome.* These diseases can be directly attributed to something gone awry in the genetic code or distribution. They are the result of defective genes or gene distribution.

As stated earlier, DNA—the only living material capable of replication—makes up the genes found on the chromosomes. It is DNA that controls the process of heredity. DNA directs the manufacture of amino acids into the proteins that are essential for normal body operation.

Thus DNA has two major functions: DNA transcription and protein synthesis. After DNA replicates itself during mitosis, DNA transcribes itself into RNA, which then goes on to protein synthesis.

RNA is the other nucleic acid, and contains ribose sugar, rather than deoxyribose sugar, in its nucleotides. (Remember that a nucleotide is composed of a phosphate, sugar, and a base. In RNA the base uracil replaces the thymine found in DNA. RNA, unlike DNA, is not structured as a double helix but is a single coil.)

There are three types of RNA, all made from the chromosomes within the nucleus of the cell, and all of them move out of the nucleus into the cytoplasm of the cell. The three types of RNA are *messenger RNA, ribosomal RNA,* and *transfer RNA.*

Protein synthesis takes place in the cytoplasm of the cell. As a gene becomes available for transcription, the helix unwinds. The messenger RNA is transcribed on the DNA template in the nucleus. This messenger RNA corresponds in length and information to the activated gene. The messenger RNA now moves from the nucleus to the cytoplasm, and attaches to the ribosome.

Ribosomal RNA is in the ribosomes, which are made up of protein and ribosomal RNA. Transfer RNAs are always present in the cytoplasm, and are specifically designed for one of twenty different types of amino acids. The transfer RNAs serve as carriers to bring amino acids to the ribosome.

The transfer RNAs bring the correct *amino acid* to the ribosome, with messenger RNA attached. The amino acids are then arranged in the proper sequence, according to the message carried by messenger RNA. A peptide bond is formed between each amino acid brought to the ribosome, and the transfer RNA is released.

Bonded amino acids form a peptide chain of a protein molecule. This continues until a polypeptide chain is completed and the messenger RNA is completely translated into protein, forming a new protein.

The precise sequence of nucleotides in DNA always determines the assembly of amino acids into proteins. A mistake can occur in the amino acid sequence, and an incorrect polypeptide chain and protein may be synthesized. The wrong amino acid can be translated.

For example, it has recently been discovered that a genetic disorder may trigger manic depression. Due to an accident in transcription, certain people develop an abnormal sensitivity to a normal brain chemical called *acetylcholine.* Their biochemistry gives such sensitized persons a greater tendency toward depression.

Cancer is also thought by many researchers to be genetic in origin. There is strong evidence, through recent scientific breakthroughs, that specific genes are closely linked to certain strains of cancer. These genes are called *oncogenes.*

A type of oncogene called a proto-oncogene is found in most normal cells. It is hypothesized that these proto-oncogenes play an important role in cell division and differentiation. But when something goes wrong with these proto-oncogenes, they become malignant. Researchers are now attempting to discover why this happens, and what can be done to prevent this transmutation. Again, there is evidence that something goes wrong in transcription.

Many institutions today are working on genetic answers to many of the diseases that plague the human condition. Some answers are already in practice, such as genetic counseling. In this field, experts work with families that are at high risk for genetic disorders. This is a communication process, in which the risk of genetic disorder is studied (through family histories and individual testing), and the family is counseled. The process helps the family to understand the specific disorder, the way inheritance contributes to the problem, and the risk to specific individuals, their options, and perhaps treatment. Genetic counseling helps the family to decide on a course of action best suited to their needs.

Other solutions to genetic disease are also being sought. Some scientists are working on vaccines that can alter the DNA, while others are working on altering the genes in the sperm and egg cells. This research is still in its infancy; and at present prevention, through counseling, seems to be the best method for combating genetic disorders.

MEDICAL GENETIC PROJECTS AT FAIRS

Do a project on the various forms of genetic research being carried on today. Report on these efforts, and postulate their future and impact on the human condition. Show how these efforts may be a major turning point for humanity. Project what medicine might be like in the year 2015.

Present a project on genetic counseling. Demonstrate how genetic counseling operates. List its benefits and drawbacks. Prepare several genetic profiles, and display them in your project booth. Perform genetic profiles on judges and spectators at the fair.

GLOSSARY

ACETYLCHOLINE. A chemical found in the human brain and at many nerve endings.

ACQUIRED CHARACTERISTICS. Characteristics that are developed through use or lost through disuse—for example, calluses on the hands.

ADENINE. A nitrogenous base found in DNA.

ALLELES. Genes that occupy corresponding positions on homologous chromosomes.

AMINO ACID. An organic compound that is the structural unit of a protein molecule.

AMNIOTIC FLUID. The fluid that surrounds a developing embryo and fetus in human beings and many other vertebrates.

AMPHIBIA. A class of vertebrates that includes frogs, toads, salamanders, and newts.

ANAPHASE. A stage in mitosis and meiosis, during which duplicate chromosomes move to opposite poles of the cell.

ANGIOSPERMS. Plants with flowers that produce seeds. These seeds and tissues, in which they are enclosed, constitute a fruit.

ASEXUAL REPRODUCTION. A process in which only one parent is responsible for the production of an offspring.

ASTERS. Fibers that form star-shaped structures around the centrosome during the prophase of mitosis.

AUTOSOMES. The twenty-two pairs of chromosomes in each cell of the human body that are not sex chromosomes.

AVES. A class of vertebrates that includes birds.

BINARY FISSION. A method of asexual reproduction in which the organism divides into two equal parts.

BUDDING. A method of asexual reproduction in which an organism divides into two unequal parts.

CALYX. In a flower, the entire circle of sepals.

CAMBIUM. A tissue that enables plants to grow wider and taller.

CELL. The smallest unit of living matter. All living things are made up of cells.

CELL MEMBRANE. A thin layer of protoplasm that bounds the cell. It is permeable, and is found in both plant and animal cells.

CELL THEORY. All living things are composed of cells that carry on their life function. All cells are derived from preexisting cells.

CENTRIOLE. An organelle of an animal cell that is associated with spindle fibers in mitosis and meiosis.

CENTROMERE. Region on a chromosome that holds sister chromatids together and attaches to spindle fibers.

CENTROSOME. A chromatin thread or bar.

CERVIX. The lower, narrow end of the uterus that opens into the vagina.

CHROMATIN. Nuclear material that condenses to chromosomes during mitosis.

CHROMOSOME. Rodlike structure that becomes visible in cells during mitosis and meiosis. Chromosomes consist of DNA and contain genes, which are responsible for heredity.

CILIA. Hairlike projections lining many passageways in the human body.

CLASSIFYING. Part of the scientific method. Putting objects into groups.

CONCLUSION. Part of the scientific method. Was your hypothesis right or wrong? Why or why not?

CONTROLLED EXPERIMENT. An experiment in which the researcher controls the variables.

COROLLA. In a flower, the entire circle of petals.

CROSS-POLLINATION. The transfer of pollen grains from the anther of one plant to the stigma of another.

CROSSING OVER. The exchange of genetic material between homologous chromosomes during meiosis.

CYTOPLASM. The liquid in a cell surrounding its nucleus.

CYTOPLASMIC DIVISION (CYTOKINESIS). The division of the liquid in the cell during telophase.

CYTOSINE. A nitrogenous base found in DNA.

DARWIN, CHARLES (1809–82). Pioneer in the science of evolution. Wrote *The Origin of Species by Means of Natural Selection.*

DAUGHTER CELLS. The two new cells resulting from mitosis and meiosis.

DEOXYRIBONUCLEIC ACID (DNA). The substance that makes up genetic material (chromosomes and genes). It is the only living substance on Earth capable of replication.

DE VRIES, HUGO (1848–1935). Dutch botanist who developed mutation theory of evolution.

DIHYBRID CROSS. A cross between organisms that are hybrid for two traits.

DIPLOID NUMBER. The number of chromosomes in the body cells of an organism.

DOMINANT CHARACTER. The character that will always show up in first-generation offspring.

DOWN'S SYNDROME. A genetically-related disease characterized by a small flattened skull, slanting eyes, and mental retardation.

DROSOPHILA MELANOGASTER. Fruit fly.

EJACULATION. Discharge of seminal fluid from the penis.

EMBRYO. An organism in its beginning stages of development.

ENDOCRINE GLAND. All the glands in the body without tubes (ducts). Endocrine glands release a liquid directly into the bloodstream.

ENDOPLASMIC RETICULUM. An organelle found in the cytoplasm, composed of a system of interconnected membranous tubules and vesicles.

ENZYME. A protein, synthesized by a cell, acting as a catalyst in specific cellular reactions.

EPIDIDYMIS. Coiled tubules, crowning the testes, that lead from the testes to the vasa deferentia.

ESTROGEN. The female hormone.

EUGENICS. A way to improve the human condition through improving heredity.

EUTHENICS. A way to improve the human condition through improving the environment.

EVOLUTION. The process of change in living things.

F_1 GENERATION. First-generation offspring.

F_2 GENERATION. Second-generation offspring.

FALLOPIAN TUBE. The tube that carries the ova from the ovary to the uterus.

FERTILIZATION. The union of egg and sperm cells.

FILAMENT. In plants, a thin stalk supporting the anther.

FLAGELLUM. A whiplike tail on sperm cells, used for propulsion.

GAMETES. Either sex cell, sperm or egg.

GAMETOGENESIS. The production of gametes.

GENE. A portion of DNA that carries the hereditary traits.

GENETIC COUNSELING. A communication process, which provides genetic information to families to help them make reproduction decisions.

GENETICS. The study of heredity; how traits are passed on from one generation to the next.

GENOTYPE. The combination of genes present within the cell of an individual.

GLAND. A group of tissues working together to perform the same function. All glands produce and secrete a liquid (hormone or enzyme).

GONADS. The ovaries or the testes.

GUANINE. A nitrogenous base found in DNA.

HAPLOID NUMBER. Half the diploid number of chromosomes.

HEREDITY. The transmission, or passing, of genetic information from one generation to the next.

HERMAPHRODITE. One individual that carries both the male and female reproductive apparatuses.

HETEROZYGOUS. An organism that possesses two different alleles for the same trait.

HOMOLOGUE. One of a matched pair of chromosomes.

HOMOZYGOUS. An organism with like alleles in a gene pair.

HORMONE. A substance secreted by endocrine glands.

HYBRID. An organism with both dominant and recessive characters, in which only the dominant manifests itself.

HYPOTHESIS. Part of the scientific method—an inference or prediction that can be tested.

IMMUNE SYSTEM. The system that combats disease in the human body.

INCOMPLETE DOMINANCE. Where neither parent's gene shows dominant in the offspring but is, rather, codominant.

INDEPENDENT ASSORTMENT (Mendel's Law of). Where each character behaves as a unit, and is inherited independently of any other character.

INFERENCE. Part of the scientific method—an educated guess based upon something you have observed about something that has happened.

INHERITANCE. The genetic characteristics received by the offspring from the parents.

INTERPHASE. Period between cell division (time of maximum growth), during which the cell carries on its normal functions.

LARVA. One of the stages in insect metamorphosis.

LAW OF DOMINANCE. One of Mendel's laws: When two organisms, pure for contrasting characters, are crossed, the dominant character will always manifest itself in the hybrid, and the recessive character will be hidden.

LAW OF SEGREGATION AND RECOMBINATION. One of Mendel's laws: When hybrid organisms are crossed, the recessive character is separated from the dominant character. At fertilization, there is a chance these recessive characters will recombine in the F_2 generation.

LINKAGE. The alignment of genes in a specific order on the chromosome.

MAMMALIA. A class of vertebrates, including human beings.

MEIOSIS (reduction division). Cell division by which egg and sperm cells are formed.

MENSTRUATION. Loss of blood and tissue from uterus at end of monthly menstrual cycle.

MERISTEM. The growth tissue in plants.

MESSENGER RNA. A molecule of RNA that transmits the information for protein synthesis from the nucleus to the cytoplasm.

METAPHASE. A stage of mitosis and meiosis, in which the chromosomes become aligned in the middle of the spindle.

MITOCHONDRIA. Organelles found in the cytoplasm that contain enzymes responsible for aerobic respiration.

MITOSIS (cell division). The process by which a cell divides into two identical daughter cells. Cell division by which somatic cells divide.

MONOHYBRID CROSS. An organism hybrid for only one trait.

MOTILE. Able to move.

MULTICELLULAR. Having more than one cell.

MUTATION. An alteration in the DNA (genetic material) of a chromosome, or a permanent change in an organism caused by such an alteration.

NONDISJUNCTION. Failure of a chromosome to separate during meiosis or mitosis.

NUCLEAR MEMBRANE. A thin layer of material that bounds the nucleus, is selectively permeable, and separates the nucleus from the cytoplasm.

NUCLEOLUS (little nucleus). A small sphere in the nucleus that contains RNA.

NUCLEUS. A spherical body, separated from the cytoplasm of a cell, that contains DNA and RNA.

OBSERVATION. Part of the scientific method—using all your senses to find out all you can.

ONCOGENES. Any genes that have the ability to start a cancer.

ONCOLOGY. The study of tumors.

OOGENESIS. The process in which an egg cell develops.

ORGAN. A group of tissues working together to perform the same function.

ORGANELLE. A part of a cell that performs a specific function.

ORGANISM. Any living thing.

OVARIES. The female reproductive glands; produce estrogen and ova.

OVULATION. The release of a mature egg cell from the ovary, and its journey through the Fallopian tube to the uterus.

OVULES. Round bodies within a plant ovary.

OVUM. The egg cell.

PARTHENOGENESIS. The development of a new offspring from an egg that has not been fertilized.

PEDIGREE. A history of genetic information.

PENIS. The male reproductive organ.

PEPTIDE. A compound composed of two or more amino acid molecules joined together chemically.

PETALS. Brightly colored leaflike structures of a flower.

PHENOTYPE. The physical manifestation (appearance) of an individual due to a particular combination of a set of genes.

PISCES. A class of vertebrates, including fish.

PISTIL. The female reproductive organ of a plant.

PITUITARY GLAND. The master endocrine gland, located at the base of the brain. It controls the functions of other endocrine glands.

PLACENTA. A structure to which a developing organism is attached, and through which it is nourished.

PLASTIDS. Bodies that float in the liquid of the cell.

POLLINATION. The transfer of pollen grains from an anther to a stigma.

POLYPEPTIDE CHAIN. A compound formed by the union of amino acid molecules.

PREDICTION. Part of the scientific method—an educated guess, based on what you have observed, about something that is going to happen.

PROPHASE. The first stage of mitosis and meiosis, in which chromosomes become visible.

PROSTATE GLAND. Part of the male reproductive system that contributes seminal fluid to sperm.

PROTOPLASM. A jellylike fluid, the living substance present in all cells. (Organelles of the cell are suspended in the protoplasm.)

PUNNETT SQUARE. A diagrammatic method of recording genetic crosses.

PUPA. A stage of insect metamorphosis.

RECESSIVE CHARACTER. Any character inherited by the offspring that does not show up (unless in pure form) in the first generation.

REGENERATION. The ability of a plant or animal to replace lost parts.

REPTILIA. A class of vertebrates consisting of turtles, snakes, and lizards.

RESULTS. Part of the scientific method—what happened during your experiment.

RIBONUCLEIC ACID (RNA). A nucleic acid that contains ribose sugar, a phosphate, and a nitrogenous base.

RIBOSOMAL RNA. A molecule of RNA that functions in protein transcription.

RIBOSOME. An organelle found in the cytoplasm, made up of protein and RNA, and functioning in protein synthesis.

SCROTUM. A saclike structure that holds testes.

SELF-POLLINATION. When pollen grains from the anther fertilize the stigma of that same flower.

SEMEN. A saline liquid in which sperm swim.

SEMINAL VESICLE. Tubelike structure that coats sperm with a sugar.

SEPALS. Outermost green, leaflike structures located at the base of the flower.

SEX LINKAGE. Refers to genes located on the X chromosome that occur more frequently in one sex than the other; for example: hemophilia and color blindness.

SICKLE CELL ANEMIA. A genetically caused disease.

SPERM. The male reproductive cell.

SPERMATOGENESIS. The process by which mature sperm cells are formed.

SPONTANEOUS GENERATION. The belief that living organisms can arise from nonliving material.

SPORANGIUM. Spore case.

SPORE CASE. A special structure that produces spores.

SPORES. Sex cells for asexual reproduction.

STAMEN. The male reproductive organ of a plant.

STIGMA. Topmost, sticky portion of the pistil, designed to capture pollen grains.

STYLE. The tube that connects stigma to ovary of plant.

SYNAPSIS. In meiosis, the lining up of homologous chromosomes in pairs near the equator of the cells, after the nuclear membrane disappears.

TAY-SACHS DISEASE. A genetically inherited disease.

TELOPHASE. A stage of mitosis and meiosis, in which daughter cells become individual structures.

TESTES. The male reproductive glands. They produce sperm and testosterone.

TESTOSTERONE. The male hormone.

THYMINE. One of the nitrogenous bases that make up DNA.

TISSUE. A group of cells that perform similar functions.

TRAITS (inherited). The physical manifestations of genes. Passed from one generation to the next; for example, eye color, hair color, and blood type.

TRANSCRIPTION. Main function of DNA during protein synthesis.

TRANSFER RNA. The molecule of RNA that carries the amino acid to the ribosome during protein synthesis.

TRISOMY. Having one or a few chromosomes triploid in an otherwise diploid set.

UNICELLULAR. Any organism composed of only one cell.

URETHRA. Tube that passes urine out of the body.

URINARY BLADDER. An organ that stores urine until it is passed out of the body.

UTERUS. The saclike, muscular organ within which the fetus develops.

VACUOLE. A space or cavity within the cytoplasm of a cell.

VAGINA. Female reproductive organ.

VARIATIONS. Differences.

VASA DEFERENTIA. Tubes in which sperm are stored and then carried to the seminal vesicles.

VEGETATIVE REPRODUCTION. A form of asexual reproduction.

VERTEBRATE. Any animal with a backbone.

YOLK. Found in egg cell; nourishes the embryo.

ZYGOTE. A fertilized egg cell.

BIBLIOGRAPHY

Anderson, Bruce. *The Price of a Perfect Baby*. Minneapolis, Minn.: Bethany House, 1984.

Bodmer, W. F. and Cavalli-Sforza, L. L. *Genetics, Evolution and Man*. San Francisco: W. H. Freeman, 1976.

Dunbar, Robort E. *Heredity*. New York: Franklin Watts, 1978.

Eiseley, Loren. *Darwin's Century*. New York: Doubleday, 1958.

Fox, L. Raymond and Elliot, Paul R. *Heredity and You*. Dubuque, Iowa: Kendall-Hunt, 1983.

Gardner, Eldon J. and Mertons, Thomas R. *Genetics Laboratory Investigations*. Minneapolis, Minn.: Burgess, 1980.

Hart, Daniel. *Our Uncertain Heritage: Genetics and Human Diversity*. New York: Harper & Row, 1984.

Lygre, David. *Life Manipulation: From Test Tube Babies to Aging*. New York: Walker, 1979.

Stickborger, Monroe W. *Genetics*. New York: Macmillan, 1976.

Stine, Gerald J. *Laboratory Experiments in Genetics*. New York: Macmillan, 1973.

Von Blum, Ruth C. *Mendelian Genetics: A Problem Solving Approach*. Wentworth, NH: COMPress, 1979.

Watson, Jack E. *Introductory Genetics: A Laboratory Textbook*. Dubuque, Iowa: Kendall-Hunt, 1976.

GROUPS TO CONTACT
IN YOUR CITY

American Cancer Society
Association for Retarded Citizens
Committee to Combat Huntington's Disease
Cystic Fibrosis Foundation
March of Dimes Birth Defects Foundation
National Association for Sickle Cell Disease
Your state's Genetics Education Center

INDEX